BIOLOGICAL FOULING OF INDUSTRIAL WATER SYSTEMS: A Problem Solving Approach

Marc W. Mittelman and Gill G. Geesey

with contributions by

J.W. Costerton, Bill McCoy, Theodore H. Meltzer,
Tom Novitsky, Rip G. Rice, and Harry F. Ridgway

Water Micro Associates
San Diego, California

Water Micro Associates
P.O. Box 28848 San Diego, CA 92128-0848

ISBN 0-9619642-0-0

Table of Contents

Contributors

Joe Boivin, Ph.D., Microbios, Ltd., Calgary, Alberta, Canada

J.W. Costerton, Ph.D. Microbios Ltd., Calgary, Alberta, Canada

Gill G. Geesey, Ph.D. Associate Professor, CSULB, Long Beach, CA

Bill McCoy, Ph.D. Research Associate, Great Lakes Chemical Corp., West Lafayette, IN

Theodore H. Meltzer, Ph.D. President, Capitola Consulting Co., Arlington, VA

Marc W. Mittelman, M.S. Technical Director, Water Micro Associates, San Diego, CA

Tom Novitsky, Ph.D. Vice President, Associates of Cape Cod, Inc., Woods Hole, MA

Rip G. Rice, Ph.D. President, Rice, Inc., Ashton, MD

Harry F. Ridgway, Ph.D. Chief Research Microbiologist, Orange Co. Water District, Fountain Valley, CA

PREFACE

The use of industrial waters for various manufacturing processes has grown dramatically over the past several decades. In addition to its traditional role as a cooling fluid, water is finding increased use as a critical raw material for numerous finished and in-process materials. Products as diverse as potable water, foods, parenteral drugs, and semiconductor devices all depend upon industrial waters of varying purities as a critical component in their manufacture. Quality requirements for cooling waters used in spacecraft, nuclear power plant, microelectronic, and other high technology applications are also increasing with the new technologies.

This book addresses a very important aspect of industrial water system operations: biological fouling. Biological fouling can be defined in terms of its effects on processes and products. 1. Material degradation 2. Product contamination 3. Mechanical blockage 4. Impedence of heat transfer. The ability of bacteria and other microorganisms to utilize available nutrient sources, reproduce, and generate, *in situ*, intra- and extracellular organic and inorganic compounds distinguishes them from other industrial water system contaminants. Traditional water treatment strategies have, for the most part, failed to address those factors which promote biofouling activities. An understanding of the molecular and physiological basis for these activities is essential to any long-term control strategies; accordingly, the authors have endeavored to present a discussion of biofouling activities in this context.

Although a theoretical basis for biofouling activities is presented, much of the text is devoted towards a practical, problem solving approach. Emphasis has been placed on a practical application of the principles of environmental microbiology to industrial water systems. The genesis and interactions of surface-associated microbial populations are described throughout the book as they relate to biofouling activities. In particular, the chapters on microbially influenced corrosion, reverse osmosis membrane fouling, and purified water systems contamination rely on such a mechanistic approach.

Laboratory methods for detecting, enumerating, and identifying fouling microorganisms have traditionally relied upon techniques developed for medical microorganisms. In chapters 10-14, an attempt has been made to present a number of methods specifically designed for processing the unique types of biological samples that are often associated with industrial and environmental waters. Where possible, these methods have been presented in a step-by-step fashion, along with equipment and reagent sources, references, and illustrations.

It is hoped that this book will aid researchers and practitioners in a number of multidisciplinary fields: industrial microbiology, facilities engineering, materials sciences, contamination control, quality assurance, and related areas. Inasmuch as the impetus for developing this work was derived from our Annual Laboratory Workshops in Biological Fouling of Industrial Water Systems, we are indebted to all of the speakers and participants of past years for their valuable discussions and encouragement.

Marc W. Mittelman　　　　　　Gill G. Geesey

November, 1987

CHAPTER 1. SURVIVAL OF MICROORGANISMS IN LOW NUTRIENT WATERS

Gill G. Geesey, Department of Microbiology, California State University, Long Beach, CA 90840

Introduction

The chemical and physical characteristics of industrial waters vary greatly depending on their uses or applications. For example, high purity water used in the pharmaceutical and electronics industries has considerably lower concentrations of inorganic and organic matter than water used to cool heat exchanger surfaces. In spite of these differences, virtually all industrial waters will foster the growth of microorganisms. Regardless of the effort directed towards maintaining aseptic conditions or sterility, contaminating microorganisms eventually invade the water system and replicate to the extent that biologically-induced problems occur. Although in any given system, the conditions are likely to be sub-optimal for growth of most of the microbes that are introduced, their physiological diversity optimizes the chances for at least one cell finding the existing conditions favorable for colonization and growth. In the case of microorganisms, growth refers to an increase in the number of individuals in the system and this in turn leads to an increase in microbial activity and biomass.

Although factors such as temperature, pH, redox potential, and oxygen concentration control which particular types of microorganisms will be present, the abundance and availability of a utilizable energy source generally limits the extent to which the invading microorganisms impact a habitat [35]. Since heterotrophic bacteria are the most common microorganisms to inhabit aquatic environments, reduced organic compounds represent the limiting energy source in these systems. Amino acids, carbohydrates, and lipids are the most common sources of energy for heterotrophic bacteria. Reduced inorganic compounds such as hydrogen sulfide, ammonia, or a reduced metal ion such as ferrous iron may serve as energy sources for chemolithotrophic bacteria that become established in a system. Besides an energy source, additional nutrients such as nitrogen, phosphorus, and sulfur plus vitamins and trace elements are needed for microbial growth. Like an energy source, these other nutrients can also limit the growth of microbes if they are in short supply.

When growth-promoting resources are not available in sufficient concentrations in the bulk water phase to support cell reproduction, microorganisms employ a variety of strategies to optimize their survival until conditions become more favorable. Some of the strategies appear to be utilized by a number of microbial species over a range of conditions while others are only infrequently encountered. It should be pointed out that many survival mechanisms are not expressed when microorganisms are subjected to the laboratory conditions used most commonly for microbiological research. Only after selective pressures are applied, such as energy removal, do the strategies become evident. Since most aquatic habitats are energy limiting, the strategies employed for survival are likely to be in effect more often than not and therefore must be considered important adaptations that have evolved to ensure maintenance of the species. This chapter describes some of the survival strategies likely to be used by heterotrophic microorganisms that exist in energy-limited aquatic habitats. Many of these strategies are undoubtedly employed by microorganisms that inhabit industrial water systems.

Low Nutrient Survival

Survival Strategies

Studies with marine bacteria have provided considerable insight into the survival mechanisms that are employed by aquatic microogranisms. Seawater generally contains organic carbon concentrations of <1 to 70 mg/l, a range which includes the range of total organic carbon (TOC) concentrations associated with many industrial water systems. Subsurface waters in the open ocean generally contain <1 mg TOC/l. Bacteria in industrial waters are thus likely to be exposed to similar energy limitations as marine bacteria; the only major differences being the greater availability of surfaces and possibly shorter periods between energy replenishment in the former.

The ability of bacteria to grow in a medium containing low concentrations of organic carbon is a characteristic that has been recognized for many years. Zobell and Grant [55] showed that *Escherichia coli*, *Staphylococcus citreus*, *Bacillus megatherium*, *Proteus vulgaris*, and *Lactobacillus lactis* multiplied in solutions containing 0.1 mg of glucose which represents 40 μg C/l. More recently, *Pseudomonas aeruginosa* has been shown to grow in tap water at 25 μg C/l supplied by any one of a variety of compounds [50]. *Aeromonas hydrophila* multiplied in a medium that contained only 10 μg C/l supplied as glucose [49]. The threshold concentration for glucose to stimilate growth of this bacterium was approximately 2.5 μg/liter. These levels of growth-stimulating nutrients are so low that they are often below the detection limits of the TOC analyzer. Schmidt and Alexander [43] found that even when the energy source was below the threshold concentration for growth stimulation, the compound could still be taken up and metabolized by the bacteria when higher concentrations of other organic molecules were present. Theoretical arguments have been presented which indicate that bacterial growth does not occur if the sole or combined energy source is at concentrations below 1 ng/l [44]. We do not know the threshold concentrations of various nutrients required by bacteria for maintenance energy or that just needed for survival of the organism. It is not surprising then that problems associated with microbial growth occur in even high purity industrial waters which generally contain 5-100 μg TOC/l.

It has been demonstrated that airborne organic contaminants support the growth of aquatic bacteria. Geller [15] showed that volatile organic compounds contributed from 0.2 to 0.5 mg dissolved organic carbon (DOC) per liter of distilled water per week, depending on the type of closure that partitioned the liquid phase from the outside atmosphere. Growth-supporting, airborne contaminants were shown to penetrate air traps filled with activated carbon and cotton wool. Growth rates of bacteria were proportional to the invasion rates of the airborne organic substances. A 10-fold increase in bacterial numbers was achieved within periods of from 1 to 8 weeks. It was estimated that 0.5 mg airborne DOC/l yielded >4.2 x 10^6 bacteria. Thus, aquatic bacteria appear to be very efficient in converting even airborne organic compounds into cell biomass.

It has been proposed that many bacteria actually prefer energy to be present in very low concentrations. Ishida *et al*. [23] showed that the maximum number of bacteria produced per unit amount of energy consumed was greater when exogenous energy levels were low. The data indicated that cells became more efficient at producing biomass as available energy decreased.

The term "oligotroph" was coined to describe those bacteria that grow in a medium containing <1 mg/l of organic carbon supplied as a complex mixture of compounds in the form of peptone, Tripticase, and yeast extract [22]. It was indicated that many oligotrophs do not grow in medium containing organic carbon concentrations of 5 g Tripticase/l and that these characteristics distinguish them from copiotrophic bacteria which require relatively high concentrations of organic carbon for growth. Martin and MacLeod [34], however, cautioned against such classification criteria and presented evidence that each bacterium possesses its own threshold concentrations for each organic compound that serves as an energy source and that too much or too little will result in growth inhibition.

Novitsky and Morita [40,41] demonstrated that when cells of a marine vibrio were transferred to a medium lacking an exogenous energy source, their initial response was to utilize endogenous energy reserves to complete rounds of DNA replication and cell division. This "fragmentation" process led to an increase in the total number of bacteria in the system during the first few days of starvation. This type of physiological response to energy limitation has been noted in a number of marine bacteria and appears to reflect a strategy in which the more individuals that exist in the population, the better chance there is for one to survive. When energy is again available, a single bacterium has the potential to replicate and reestablish the population.

During the period of cell fragmentation, respiration has been reported to be high [27]. Zimmerman _et al_. [53] showed that 1% of a starving bacterial population exhibits respiratory activity. The overall respiratory level of the total bacterial population, however, usually declines during starvation. Novitsky and Morita [40] demonstrated that overall respiration was reduced by 99% after 1 week of starvation. Interestingly, cells starved for 2 weeks had more ATP per viable cell than unstarved cells [2,29]. Whether reduction of respiratory activity and ATP conservation represent a survival mechanism remains to be determined.

Bacteria appear to pass more deeply into dormancy with longer starvation times. Amy *et al*. [2] showed that bacteria starved for longer periods of time exhibited longer lag periods before exponential growth resumed following addition of an energy source. This was suggested to result from delays in macromolecular synthesis. No delays were noted, however, in the uptake of glutamic acid by starved cells [1]. It was postulated that exposure to a sufficient supply of energy may be necessary before metabolism of nutrients taken up by the cell is directed toward processes leading to cell replication, and that this may represent yet another strategy for survival of bacteria in low energy environments.

After a period of energy limitation, only a small portion (0.1%) of the initial cell population appears viable on the basis of the ability of cells to reproduce when energy is provided [29]. Nevertheless, 10^5 bacteria/ml remained viable after a 1 year starvation period. Some starved cells remained viable for as long as 2.5 years [35].

When subjected to energy depletion, many marine bacteria have been shown to reduce their cell volume. "Ultramicrobacteria" with a cell diameter less than 0.3 μm replace larger cells in laboratory cultures subjected to starvation [39]. Tabor *et al*. [45] showed that a large portion of the bacteria recovered from the deep ocean were small enough to pass through a filter with a pore size of 0.45 μm. *Alcaligenes*, *Flavobacterium*, *Pseudomonas*, and *Vibrio* were among the bacterial types that were recovered in the sample filtrates. These filterable bacteria increased in size when cultured in laboratory media [46,31]. The abundance and nutritional requirements of ultramicrobacteria imply that they represent a class of microorganisms that have successfully adjusted to low nutrient conditions. Reduction of cell size increases the surface area/volume ratio which may increase the concentration of nutrients scavenged from the environment inside the cell to levels necessary for enzyme activity.

Reduction of cell diameters to 0.3 μm or less raises questions concerning the accuracy of enumeration techniques that employ filtration methodology and direct light microscopic examination to estimate bacterial densities in low-nutrient waters. Since the resolution of most light microscopes is on the order of 0.2 μm, objects in this size range are difficult to resolve. It may therefore be necessary to use electron microscopy as a method of direct enumeration in water systems containing significant numbers of ultramicrobacteria.

Not all starved bacteria exhibit fragmentation and size reduction. Morita [37] described situations in which no changes in cell concentration and no reductions in cell volume occurred with the onset of starvation. Ishida and Kadota [22] indicated that no change in cell size occurred over a wide range of nutrient conditions and Jones and Morita [25] reported that cell size remained constant during starvation of marine nitrifying bacteria.

Starvation has been reported to induce more subtle changes in bacterial physiology and morphology than the obvious changes discussed above. Kjelleberg and Hermansson [26] described changes in cell surface charge and hydrophobicity during starvation of several marine isolates. Of 7 strains examined, 4 exhibited an increase in hydrophobicity. This physiochemical change was accompanied by an increased roughness of the cell surface throughout the starvation period.

During starvation survival, the ability to scavenge energy-yielding nutrients from the bulk medium is enhanced in some bacteria. Starved rods and coccoid cells of an aquatic Cytophaga sp. were able to take up glucose at lower external concentrations than unstarved cells [42]. Likewise, glucose respiration and incorporation into macromolecules was demonstrated to occur at lower exogenous glucose concentrations in starved cells than in actively-growing or non-viable cells.

High affinity uptake systems are used by bacterial cells to scavenge those utilizable, energy-yielding, organic substances that are present in very low concentrations in the surrounding medium. The marine bacterium, Ant-300, possesses two uptake components for the amino acid arginine; one with high affinity that operates in the nanomolar range and a lower affinity component that operates in the micromolar range [13]. High affinity systems have also been described in bacteria that scavenge ammonia from the atmosphere [24]. It is possible that as higher quality water is produced for industrial applications, microorganisms will evolve to survive and grow on lower concentrations of nutrients.

Uptake systems appear to be maintained by the bacteria even when the compounds they transport are too dilute to be taken up into the cell. Mannitol transport by a marine pseudomonad was depressed during the first 30 h of starvation but could be readily reinduced by exogenously supplied mannitol after 6 weeks [9]. Seventy two-day starved cells of Ant-300, when presented with 10^{-12} M glutamic acid, utilized the amino acid without any lag period [36]. Thus, the genetic machinery and an endogenous energy reserve for nutrient uptake into the cell is maintained for long periods of time in starved bacteria.

Chemotaxis is a response by microorganisms to a chemical concentration gradient. Many motile bacteria display a series of self-propelled movements which direct them to an area where a nutrient or energy source is in higher concentration. Geesey and Morita [13] showed that cells of the marine bacterium, Ant-300, were nonmotile in an energy-rich medium but after 48 hours of starvation, became motile and elicited a positive chemotactic response toward the amino acid arginine. The threshold concentration of arginine required to stimulate a chemotactic response by this bacterium was 10^{-7} M [47]. Starvation periods longer than 48 h appeared to inhibit chemotaxis in Ant-300. Chemotactic responses are probably not important in turbulent flow conditions but may be significant in transporting bacteria through the viscous sublayer adjacent to a submerged surface [5,6].

A characteristic of virtually every industrial water system is the presence of surfaces. "Inert" surfaces have long been recognized as sites of nutrient enrichment [54]. Organic films accumulate on the cleanest surfaces within hours of exposure to water containing dissolved organic compounds. The concentrations of dissolved organic carbon need not be high for surface adsorption to occur. Organics in ultrapure water deposit on the inner surfaces of the distribution piping, filter media, and storage tank walls. Geller [15] presented evidence indicating that 30% of the 2.2 mg airborne carbon that entered a liter volume of distilled water subsequently adsorbed to the vessel wall. DePalma and Baier [11] used infrared spectroscopic techniques to elucidate the chemical nature of the organic matter that adsorbed to submerged surfaces and concluded that they are dominated by glycoprotein-like compounds. There is growing evidence that, in systems such as those circulating industrial water, surfaces possess higher concentrations of energy-yielding organic molecules than does the bulk aqueous phase.

It is not surprising then that attachment of cells to surfaces is one of the most common strategies for bacterial survival in environments where the bulk water phase is energy-limiting. Marshall _et_ _al_. [13] reported that ultramicrobacteria were primary colonizers of glass surfaces immersed in the marine environment. These bacteria were replaced by normal sized bacteria after 12-24 h. Dawson _et_ _al_. [10] demonstrated that cells deprived of nutrients become more adhesive. They suggested that adhesion is a survival tactic of copiotrophic bacteria under low-nutrient stress. Copiotrophic bacteria differ from oligotrophic bacteria in that they demand a flux of nutrients many times higher than that of the latter for similar energy-demanding processes.

Some small, starved copiotrophic bacteria exhibit changes in morphology and cell surface physiochemical properties that likely influence adhesion. Besides the increased cell surface roughness and decreased polarity of surface groups mentioned above, these bacteria also exhibit a high degree of irreversible binding to surfaces. Many of the cells that became irreversibly bound to glass beads coated with stearic acid exhibited the ability to utilize the surface-bound substrate [17]. Even the portion of attached cells that were reversibly bound to the beads appeared to have utilized some of the surface-bound fatty acid during their contact time with the beads. Thus, adherent bacteria have the ability to utilize surface-associated nutrients.

Bacteria attached to surfaces exhibit different nutrient uptake characteristics than freely-suspended bacteria. Natural populations of attached bacteria transported ATP 100 times faster than free-living bacteria on a per cell basis [18]. This was attributed to the larger average cell volumes of the attached bacteria. Geesey *et al*. [14] showed that attached cells of *Enterobacter cloacae* transported 2-7 time more glucose than freely-suspended cells at added glucose concentrations of 1.25 ng/ml.

Attachment to surfaces does not necessarily imply that more efficient transport of nutrients into the cell will occur. Bright and Fletcher [3] found that the maximum velocity of leucine uptake by a marine *Pseudomonas* sp. was similar regardless of whether the cells maintained a sessile or free-living existence. Nevertheless, assimilation of amino acids was generally greater and respiration less in attached cells than in free-living cells.

Low Nutrient Survival

Humphrey *et al.* [20] postulated that surfaces trigger a burst of cell activity at the expense of endogenous energy reserves. This resulted in an optimal survival configuration, which included a reduction in cell size. Humphrey and Marshall [19] showed that surfactants located at surfaces specifically stimulate this surface-attached bacterial activity. Protein added at very low concentrations was degraded more rapidly by bacteria on surfaces than by bacteria in the bulk liquid [16]. At higher protein concentrations, degradation rates were similar among bulk phase and surface-associated bacterial populations. Thus, when nutrients become limiting in the bulk liquid, the surface-associated bacteria appear to be the more active population in the system. In view of the conditions that characterize most industrial water systems, it is likely that the adherent bacteria will have a greater impact than the free-floating or planktonic population.

An expression of the metabolic activity of bacteria that contact a surface is the production of variable amounts of polymers that are bound to the exterior of the cell envelope [8]. ZoBell [54] was the first to suggest that extracellular polymers were responsible for the firm adhesion of bacteria to surfaces. Ulinger and White [48] suggested that extracellular polymers were synthesized by bacteria under stress. These exopolymers are generally polysaccharide in nature, but contain proteins and other cellular products, as well.

Adhesiveness of bacterial cells to surfaces has been related to the amount of exopolymer produced by the cell. Nelson *et al.* [38] suggested that the increased rate of adhesion of a *Pseudomonas* sp. was due to increased amounts of exopolymer present. Fletcher [12] indicated, however, that it is not the quantity of extracellular polymer that determines attachment but its adsorption and solubility in the system.

Other studies have shown that the synthesis of extracellular polymers does not necessarily enhance bacterial adhesiveness. Nitrogen-limited cultures exhibited little attachment to surfaces even though there was abundant exopolymer production [4,51]. A marine Pseudomonas sp. produced an extracellular polysaccharide capsule during energy starvation that showed a lower degree of adhesion to hydrophobic surfaces than did cells devoid of the polymer [52]. In fact, excessive polymer production may cause dispersion of adherent bacteria into the bulk fluid. This may occur when surface area rather than energy becomes limiting. It is not yet clear how extracellular polymers contribute to the adhesion event. The bulk of evidence at this time suggests that these polymers are important in the final stages of adhesion since they usually appear after the cells have made initial contact with a solid surface.

Extracellular polymer matrices are the most frequently observed structures anchoring natural aquatic bacterial populations to surfaces. These polymeric molecules appear to interact with the organic molecules adsorbed to the surface rather than with the actual substratum surface. In some instances, physical removal of the surface-associated bacteria also results in the removal of the underlying conditioning film (Geesey, unpublished results). Some surface-bound molecules appear to inhibit bacterial adhesion, however, through steric exclusion effects. It should be pointed out that no surface coatings have yet been formulated that effect long-term exclusion of microorganism. Marshall [32] indicates that there is much to be learned about the effects of natural conditioning films on the adhesion of and competition between bacteria colonizing a surface.

The extracellular polymers also extend between the surface-bound cells, forming a matrix that eventually develops into a biofilm [7]. The biofilm traps additional bacteria and nutrients from the bulk water phase. A consortium of different physiological types of bacteria form microniches within the biofilm. Often, conditions within the biofilm are sufficiently different from the bulk liquid phase that bacteria unable to grow in the latter, are still able to grow in the system. Metabolic by-products of one bacterium become the nutrients and energy source for another. The polymer network surrounding the bacteria maximizes retention of the bacterial products in the biofilm. Flocculation has been shown to be a microbial response to substrate limitation. Hunt and Logan [21] suggested that transport of nutrients is enhanced in the highly porous exopolymer matrix which mediates floc formation. Exopolymers appear to play a major role in the aquisition of energy for bacteria in aquatic environments.

In addition to anchoring cells to surfaces, extracellular polymers promote cell survival in other ways. The formation of microcolonies by polymer bridging between dividing cells provides protection against predators which feed on single-cells. The diffusion barrier that the exopolymers create between the bacterium and the external environment also affords some protection against bacterial viruses and toxic chemicals in the bulk liquid phase. _Legionella pneumophilia_ exhibited much greater resistance to chlorine when maintained under starvation conditions than during periods of energy-excess [28]. This was likely due to differences in the quantity or chemical nature of the capsular polymers produced under these different conditions.

The rate of bacterial transport to surfaces has been shown to be directly proportional to the concentration of bacteria in the bulk aqueous phase at low cell concentrations [5]. The rate of bacterial adsorption to a surface was also shown to depend on the specific growth rate of the bacterial population [38]. Faster growing cells adhered less efficiently.

A decrease in sticking efficiency of fast-growing bacteria transported to the surface was observed with increasing turbulence. In another study, it was shown that the initial rate of deposition of slow growing bacterial cells on solid surfaces increased with the flow rate of the bulk fluid [30]. Slower growing cells also had a longer residence time on the surface. In general, it appears that suboptimal conditions in the bulk fluid stimulate surface attachment by bacteria. Hydrophobic bacteria appear to adhere more firmly to solid surfaces than hydrophilic bacteria [27]. However, surface properties also influence the rate of bacterial adsorption.

That bacteria reversibly adhere to surfaces provides a mechanism for the dissemination of active microorganisms throughout a water system. As bacterial cells accumulate on a surface through attachment and growth, conditions for survival eventually deteriorate to the extent that detachment is favored, allowing the bacteria to colonize more suitable surface sites downstream. Given enough time, this sequence of attachment, growth, detachment, and recolonization will lead to the dissemination of bacteria throughout a system--even in one that contains very low concentrations of nutrients in the bulk water phase.

Concluding Remarks

Nutrient levels and energy availability for microorganisms in industrial waters are generally within the ranges found in natural marine and freshwater environments. Most of the contaminating microorganisms that establish populations in industrial water systems are derived from these other aquatic habitats. The contaminating microorganisms survive and in some cases flourish in these systems because they have evolved mechanisms that allow their essential biological processes to operate under energy-limiting conditions.

Studies with freshwater and marine bacterial isolates have revealed a variety of adaptations that enable microbes to survive in nutrient-poor water for extended periods of time. Adhesion to surfaces appears to be a widespread adaptation in systems containing a high surface area:volume ratio. Exopolymer production by surface-associated bacteria promotes the development of biofilms. Conditions within biofilms may differ considerably from those of the bulk water phase; microbial species poorly-adapted to conditions in either phase still have the opportunity to establish themselves in the system.

In the absence of surfaces for attachment, free-living, planktonic microorganisms survive starvation conditions by adopting strategies which include a reduction in cell volume and the utilization of high affinity uptake components to transport the low levels of energy-yielding compounds from the surrounding water into the cell. Once inside the cell, metabolism of these compounds is efficiently directed in order to sustain only the most essential cellular processes. This allows the microorganisms to pass into a state of dormancy until conditions become favorable for cell growth. In industrial water systems, this may occur as a result of changes in source water quality or from exhaustion of in-line water treatment devices. Thus, the colonization and survival strategies of microorganisms pose unique challenges to water quality management.

Monitoring microbial contamination in industrial water systems requires methodologies that are different from those used to enumerate microorganisms maintained under laboratory conditions. Assays for surface-associated populations, ultramicrobacteria, and oligotrophic bacteria must be employed in order to obtain an accurate assessment of the contamination levels. Of more importance to the concerns of industry, the assays must be effective in evaluating threshold or tolerance limits for microbial contaminants since in most cases it is impractical to maintain sterility throughout a system. These goals can only be achieved through a better understanding of the physiology and ecology of the microorganisms that colonize industrial water systems. Only then can microbially-related problems be avoided and efficient system operations be maintained.

References

1. Amy, P.S. and R.Y. Morita. 1983. Starvation survival patterns of sixteen freshly isolated open-ocean bacteria. Appl. Environ. Microbiol. 45:1109-1115.

2. Amy, P.S., C. Pauling and R.Y. Morita. 1983. Recovery from nutrient starvation by a marine *Vibrio* sp. Appl. Environ. Microbiol. 45:1685-1690.

3. Bright, J.J. and M. Fletcher. 1983. Amino acid assimilation and respiration by attached and free-living populations of a marine *Pseudomonas* sp. Microb. Ecol. 9:215-226.

4. Brown, C.M., D.C. Ellwood and J.R. Hunter. 1977. Growth of bacteria at surfaces: Influence of nutrient limitation. FEMS Microbiol. Lett. 1:163-166.

5. Characklis, W.G. 1981a. Bioengineering report: Fouling biofilm development: A process analysis. Biotech. Bioeng. 23:1923-1960.

6. Characklis, W.G. 1981b. Microbial fouling: A process analysis. p. 251-291. *In* E.F.C.

Somerscales and J.G. Knudsen (ed.), Fouling of heat transfer equipment, Hemisphere, Washington, D.C.

7. Costerton, J.W., K.-J. Cheng, G.G. Geesey, T.I. Ladd, J.C. Nickel, M. Dasgupta, and T.J. Marrie. 1987. Bacterial biofilms in nature and disease. Ann. Rev. Microbiol. 41:435-464.

8. Costerton, J.W., R.T. Irvin and K.-J. Cheng. 1981. The bacteral glycocalyx in nature and disease. Ann. Rev. Microbiol. 35:299-324.

9. Davis, C.L. and F.T. Robb. 1985. Maintenance of different mannitol uptake systems in oxidative marine bacteria during starvation. Appl. Environ. Microbiol. 50:743-748.

10. Dawson, M.P., B.A. Humphrey and K.C. Marshall. 1981. Adhesion: a tactic in the survival strategy of a marine vibrio during starvation. Curr. Microbiol. 6:195-199.

11. DePalma, V.A. and R.E. Baier. 1978. Microfouling of metallic and coated metallic flow surfaces in model heat exchange cells. In R. Gray (ed.), Proceedings of the OTEC biofouling and corrosion symposium, Department of Energy, Washington, D.C.

12. Fletcher, M. 1980. Adherence of marine micro-organisms to smooth surfaces. p. 345-374. In E.H. Beachley (ed.), Bacterial adherence, Chapman and Hall, New York.

13. Geesey, G.G. and R.Y. Morita. 1979. Capture and uptake of arginine at low concentrations by a marine psychrophilic bacterium. Appl. Environ.

Microbiol. 38:1092-1097.

14. Geesey, G.G., S.D. Salas and M.W. Mittelman. 1986. Effects of environmental conditions on the sessile existence of an estuarine sediment bacterium. p. 243-247. In Proceedings of the second colloquium of marine bacteriology, Centre National de la Recherche Scientifique, Brest, France.

15. Geller, A. 1983. Growth of bacteria in inorganic medium at different levels of airborne organic substances. Appl. Environ. Microbiol. 46:1258-1262.

16. Hermansson, M. and B. Dahlback. 1983. Bacterial activity at the air/water interface. Microb. Ecol. 9:317-328.

17. Hermansson, M. and K.C. Marshall. 1985. Utilization of surface localized substrate by non-adhesive marine bacteria. Microb. Ecol. 11:91-105.

18. Hodson, R.E., A.E. Maccubbin and L.R. Pomeroy. 1981. Dissolved adenosine triphosphate utilization by free-living and attached bacterioplankton. Mar. Biol. 64:43-51.

19. Humphrey, B.A. and K.C. Marshall. 1984. The triggering effect of surfaces and surfactants on heat output, oxygen consumption and size reduction of a starving marine *Vibrio*. Arch. Microbiol. 140:166-170.

20. Humphrey, B.A., S. Kjelleberg and K.C. Marshall. 1983. Responses of marine bacteria under starvation conditions at a solid-water interface. Appl. Environ. Microbiol. 45:43-47.

21. Hunt, J.R. and B.E. Logan. 1984. Bioflocculation as a microbial response to substrate limitations. In Proceedings of the Division of Microbial and Biochemical Technology, ACS Meeting, Philadelphia, PA.

22. Ishida, Y. and H. Kadota. 1981. Growth patterns and substrate requirements of naturally occurring obligate oligotrophs. Microb. Ecol. 7:123-130.

23. Ishida, Y., I. Imai, T. Miyagaki and H. Kadota. 1982. Growth and uptake kinetics of a facultatively oligotrophic bacterium at low nutrient concentrations. Microb. Ecol. 8:23-32.

24. Jones, K.L. and M.E. Rhodes-Roberts. 1980. Physiological properties of nitrogen-scavenging bacteria from the marine environment. Appl. Bacteriol. 49:421-433.

25. Jones, R.D. and R.Y. Morita. 1985. Survival of a marine ammonium oxidizer under energy-source deprivation. Mar. Ecol. Prog. Ser. 26:175-179.

26. Kjelleberg, S. and M. Hermansson. 1984. Starvation-induced effects on bacterial surface characteristics. Appl. Environ. Microbiol. 48:497-503.

27. Kjelleberg, S., B. Humphrey and K.C. Marshall. 1983. Initial phases of starvation and activity of bacteria at surfaces. Appl. Environ. Microbiol. 46:978-984.

28. Kuchta, J.M., S.J. States, J.E. McGlaughlin, J.H. Overmeyer, R.M. Wadowsky, A.M. McNamara,

R.S. Wolford and R.B. Yee. 1985. Enhanced chlorine resistance of tap water-adapted *Legionella pneumophila* as compared with agar medium-passaged strains. Appl. Environ. Microbiol. 50:21-26.

29. Kurath, G. and R.Y. Morita. 1983. Starvation-survival physiological studies of a marine *Pseudomonas* sp. Appl. Environ. Microbiol. 45:1206-1211.

30. Leech, R. and R.J.W. Hefford. 1980. The observation of bacterial deposition from a flowing suspension. p. 544-545. *In* R.C.W. Berkeley, J.M. Lynch, J. Melling, P.R. Rutter and B. Vincent (ed.), Microbial adhesion to surfaces, Horwood, Chichester, U.K.

31. MacDonell, M.T. and M.A. Hood. 1982. Isolation and characterization of ultramicrobacteria from a Gulf Coast estuary. Appl. Environ. Microbiol. 43:566-571.

32. Marshall, K.C. 1985. Mechanisms of bacterial adhesion at solid-water interfaces. p. 133-161. *In* M. Fletcher and D.C. Savage (ed.), Bacterial adhesion: mechanisms and physiological significance, Plenum Press, New York.

33. Marshall, K.C., R. Stout and R. Mitchell. 1971. Selective sorption of bacteria from seawater. Can. J. Microbiol. 17:1413-1416.

34. Martin, P. and R.A. MacLeod. 1984. Observations on the distinction between oligotrophic and eutrophic marine bacteria. Appl. Environ. Microbiol. 47:1017-1022.

35. Morita, R.Y. 1982. Starvation-survival of

heterotrophs in the marine environment. Adv. Microb. Ecol. 6:171-198.

36. Morita, R.Y. 1984. Substrate capture by marine heterotrophic bacteria in low nutrient waters. p. 83-100. In J.E. Hobbie and P.J.LeB. Williams (ed.), Heterotrophic activity in the sea, Plenum Publishing Corp., New York.

37. Morita, R.Y. 1985. Starvation and miniaturisation of heterotrophs, with special emphasis on maintenance of the starved viable state. p. 111-130. In M. Fletcher and G. Floodgate (ed.), Bacteria in the natural environments: The effect of nutrient conditions, Society for General Microbiology, U.K.

38. Nelson, C.H., J.A. Robinson and W.G. Characklis. 1985. Bacterial adsorption to smooth surfaces: rate, extent, and spatial pattern. Biotech. Bioeng. 27:1662-1667.

39. Novitsky, J.A. and R.Y. Morita. 1976. Morphological characterization of small cells resulting from nutrient starvation of a psychrophilic marine vibrio. Appl. Environ. Microbiol. 32:617-622.

40. Novitsky, J.A. and R.Y. Morita. 1977. Survival of a psychrophilic marine vibrio under long-term nutrient starvation. Appl. Environ. Microbiol. 33:635-641.

41. Novitsky, J.A. and R.Y. Morita. 1978. Possible strategy for the survival of marine bacteria under starvation conditions. Mar. Biol. 48:289-295.

42. Reichardt, W. and R.Y. Morita. 1982. Survival

stages of a psychrotrophic Cytophaga johnsonae strain. Can. J. Microbiol. 28:841-850.

43. Schmidt, S.K. and M. Alexander. 1985. Effects of dissolved organic carbon and second substrates on the biodegradation of organic compounds at low concentration. Appl. Environ. Microbiol. 49:822-827.

44. Schmidt, S.K., M. Alexander and M.L. Shuler. 1985. Predicting threshold concentrations of organic substrates for bacterial growth. J. Theor. Biol. 114:1-8.

45. Tabor, P.S., K. Ohwada and R.R. Colwell. 1981. Filterable marine bacteria found in the deep sea: distribution, taxonomy, and response to starvation. Microb. Ecol. 7:67-83.

46. Torrella, F. and R.Y. Morita. 1981. Microcultural study of bacterial size changes and microcolony and ultramicrocolony formation by heterotrophic bacteria in seawater. Appl. Environ. Microbiol. 41:518-527.

47. Torrella, F. and R.Y. Morita. 1982. Starvation induced morphological changes, motility and chemotaxis patterns in a psychrophilic marine vibrio. p. 45-60. In Deuxieme colloque de microbiologie marine, Publ. Centre National pour l'Exploitation des Oceans, Marseille, France.

48. Uhlinger, D.J. and D.C. White. 1983. Relationship between physiological status and formation of extracellular polysaccharide glycocalyx in Pseudomonas atlantica. Appl. Environ. Microbiol. 45:64-70.

49. Van der Kooij, D., A. Visser and W.A.M. Hijnen. 1980. Growth of Aeromonas hydrophila at low concentrations of substrates added to tap water. Appl. Environ. Microbiol. 39:1198-1204.

50. Van der Kooij, D., J.P. Oranje and W.A.M. Hijnen. 1982. Growth of Pseudomonas aeruginosa in tap water in relation to utilization of substrates at concentrations of a few micrograms per liter. Appl. Environ. Microbiol. 44:1086-1095.

51. Wardell, J.N., C.M. Brown and D.C. Ellwood. 1980. A continuous culture study of the attachment of bacteria to surfaces. p.221-230. In R.C.W. Berkeley, J.M. Lynch, J. Melling, P.R. Rutter, and B. Vincent (ed.), Microbial adhesion to surfaces, Horwood, Chichester, U.K.

52. Wrangstadh, M., P.L. Conway and S. Kjelleberg. 1986. The production and release of an extracellular polysaccharide during starvation of a marine Pseudomonas sp. and the effect thereof on adhesion. Arch. Microbiol. 145:220-227.

53. Zimmermann, R., R. Iturriaga and J. Becker-Birch. 1978. Simultaneous determination of the total number of aquatic bacteria and the number thereof involved in respiration. Appl. Environ. Microbiol. 36:926-935.

54. ZoBell, C.E. 1943. The effect of solid surfaces on bacterial activity. J. Bacteriol. 46:39-56.

55. Zobell, C.E. and C.W. Grant. 1943. Bacterial utilization of low concentrations of organic matter. J.Bacteriol. 45:555-564.

CHAPTER 2. FOULING BIOFILM FORMATION

W.F. McCoy, Great Lakes Chemical Corp., Lafayette, IN

It is generally recognized that bacteria are the initial colonizers of wetted solid surfaces in aquatic environments. Virtually all materials are susceptible to bacterial biofilm formation. The events that occur at solid-liquid interfaces prior to bacterial colonization determine the *in situ* characteristics of the surface to which bacteria adsorb. A clear understanding of these events is necessary in order to describe the formation of biofilms.

Physico-chemical Factors

For the present discussion we shall use the terms adsorbance, adsorbent and adsorbate to describe, respectively, adhesion to surfaces, the surface upon which the adhesion occurs, and the adherent molecular or colloidal material.

Adsorption of organic molecules onto surfaces

The surface characteristics of an adsorbent are instantaneously modified after exposure to natural aqueous environments [2,37,56-58]. In aqueous systems containing even very low concentrations of organic matter, this modification is caused by the adsorption of an organic film onto surfaces [6,12,14,50,88]. The reorientation of suspended organic matter in solution results in the concentration of molecules at solid-liquid, liquid-liquid or air-liquid interfaces; this effect causes a decrease in the interfacial tension. The process is thermodynamically favored because there is a loss of free energy (and an increase in entropy) when organic molecules adsorb to surfaces [13].

Many natural aqueous environments: e.g., lakes, rivers and oceans, have very low concentrations of dissolved organic material (usually in the order of μg per liter). Macromolecular adsorption to surfaces can be observed and measured even in these dilute solutions. Ellipsometry, contact angle measurements, and fluorescence spectroscopy have been used to demonstrate the development of an organic film on clean platinum plates following immersion for several minutes in Atlantic Ocean and Chesapeake Bay water [50,56,57]. The thickness of the film one hour after immersion was estimated to be less than 25 nm.

Changes in the surface "wettability" have been observed by measuring liquid contact angles before and after immersion in oceanic or estuarine waters. Figure 1 shows that the tendency of a test liquid to spread on a surface can be described by measuring its contact angle. Comparison of contact angles between drops of methylene iodide and water on clean platinum plates rinsed in organic nutrient media after immersion in oceanic or estuarine waters indicated that hydrophobic films rapidly accumulated onto the surfaces. Fluorescence spectra of naturally adsorbed organic films have shown that the organic material on surfaces immersed in oceanic and estuarine waters is very similar to humic substances [50]. Humus is the product of complex and poorly understood processes that involve the degradation of biological compounds from living organisms. Generally, humus is polyanionic and usually 1% to 5% nitrogen; 50% of the nitrogen exists in heterocyclic molecules. The rest of the nitrogen is in amino sugars, ammonia or α-amino nitrogen; up to 5% of the total nitrogen is in peptide linkages [33]. Fluorescence spectra of adsorbed organic films show that proteinaceous materials adhere to solid surfaces exposed to natural waters [50].

undirected motion: biofilm formation can be influenced by directional taxis [15,83,84]. Chemotaxis is the movement of an organism in response to a chemical gradient. Chemicals (e.g., carbohydrates and amino sugars) that illicit positive (attractive) chemotactic responses enhance the rate of bacterial attachment to artificial surfaces. Conversely, certain chemicals (e.g., $CHCl_3$ and $CuSO_4$) cause negative (repulsive) chemotactic responses which lead to active avoidance of regions with high concentrations of the chemical. Sublethal concentrations of a wide variety of toxic substances, including heavy metals and hydrocarbons, are known to cause negative chemotaxis [84]. The negative chemotactic response of certain bacteria to sublethal concentrations of toxins takes precedence even when higher concentrations of compounds that usually illicit a positive chemotactic response are present. Negative chemotaxis to sublethal concentrations of toxins was suggested as a means to control marine fouling [84]. Mathematical models of positive and negative microbial chemotaxis in flowing water have been described [46,49].

Adsorption of bacteria to solid surfaces

The scenario that we describe here is that bacteria adsorb to inert solid surfaces immersed in very dilute organic solutions. Table 1 is offered as the framework upon which we have organized this discussion of physico-chemical factors that contribute to bacterial adsorption. Any model or description of such diverse phenomena must admit many over simplifications.

The process of bacterial adsorption to solid surfaces immersed in seawater was observed by ZoBell in 1943 [88]. The first step is a reversible phase which is usually explained in terms of the electrostatic interactions between the cell surface and the adsorbent. The second step leads to a much more tenacious bond between adsorbent and adsorbate and is, therefore, generally referred to as irreversible; it is the result of the biological activity of the adsorbate.

TABLE 1. The adsorption of bacterial cells to solid surfaces immersed in very dilute organic solutions

---

Reversible Adsorption Phase.

Mechanism-Balance between double-layer repulsion and van der Waals attraction.

Net Effect-Loosely adherent cells are a finite distance (~ 2 nm) from adsorbent; cells are easily removed by rinsing.

Influencing Factors-pH, electrolyte concentration, organic carbon concentration, fluid motion.

Irreversible Adsorption Phase

Mechanism-Adhesion due to biologically-mediated adhesions (e.g., polyanionic molecules) or hydrophobicity of cell surface.

Net Effect-Tenaciously adherent cells; biologically-mediated adhesions "bridge" gap between cells and surface; cells can be removed by alteration of adhesins or by harsh physical methods (e.g., brushing or scraping).

Influencing Factors-Time, temperature, surface free energy (γ), critical surface tension (γ_c), interaction parameters (Φ), free energy of adhesion (ΔF), hydrophobicity.

Reversibly adsorbed bacterial cells are easily displaced by gentle rinsing or by fluid shear [88]. Marshall *et al*. [51] showed that reversible adsorption of bacterial cells (*Achromobacter* sp. and *Pseudomonas* sp.) to glass cover slips could be adequately explained in terms of the Derjaguin/Landau and Verwey/Overbeck (DLVO) theory [29,76]. The DVLO theory states that the distance of separation between colloidal adsorbents and adsorbates in an electrolyte solution is that distance at which the repulsive (V_R) and attractive (V_A) forces are balanced. Curtis [26,27] suggested that the repulsive forces (electrostatic double-layer) and the attractive forces (principally London dispersion force) are significant in cellular adhesion processes. The total energy of interaction (V_T) between a bacterium and a solid surface is:

$$V_T = V_R + V_A.$$

The repulsive force is the result of the electrical double-layer that exists on any charged surface in an electrolyte solution. The ionizable groups on that surface tend to attract charged groups of the opposite sign. Bacterial cell surfaces are generally polyanionic; i.e., they are negatively charged [77]. The electrostatic double-layer that exists around bacteria in an electrolytic solution consists of positively charged ions (the electrolyte) on the outer layer and negatively charged organic compounds (e.g., extracellular carbohydrates) on the inside. Consequently, there is an electrical potential across the electrostatic double-layer which is called the surface potential, ψ [38,39]. The average ψ is calculated from electrokinetic measurements because the mobility of a charged particle in an electric field (the electrophoretic mobility) is related to another potential, zeta, ζ. The outer portion of the double-layer will shear in an electric field at some radius leaving the surface with a net charge, ζ, which is proportional to the electrophoretic mobility, V:

$$\zeta = \frac{4 V \pi \eta}{\epsilon}$$

where ϵ is the dielectric constant and ξ is the viscosity of the liquid [79,80]. Under many circumstances, the electrokinetic plane of shear should coincide with the surface of the cell so that ζ and ψ are equal [38,39]. However, they may not be equal under conditions of low ionic concentration or in the presence of irregular (non-spherical) cell surfaces [27,28].

The surface potential decreases with increasing electrolyte concentration; trivalent ions lower ψ more than divalent ions and divalent ions lower ψ more than monovalent ions. There is experimental evidence relating surface potential and cellular adsorption to surfaces; these data generally show that there is an inverse correlation between adsorption and cell surface potential [27].

Marshall _et_ _al_. [51] published solutions for the calculation of the energy of interaction between glass and a _Pseudomonas_ sp. Figure 2 shows a schematic representation of the results for high electrolyte concentration (ψ is low) and for low electrolyte concentration (ψ is high). Under conditions of high electrolyte concentration there exists an attraction trough some finite distance (about 1 nm) from the adsorbent. The point at which V_R and V_A are balanced is the distance that reversibly adsorbed bacteria are held. The repulsive hump that exists closer to the adsorbent prevents a closer approach by the bacterium. The energy required to exceed this repulsion hump is greater than the motive force of the _Pseudomonas_ sp. [19]. No net attraction trough could be demonstrated under conditions of low electrolyte concentration and there were no reversibly-adsorbed bacteria observed experimentally. This work suggests that reversible adhesion to surfaces in very pure water may be insignificant. Figure 3 illustrates reversible bacterial adsorption in a solution of high electrolyte concentration (>10^{-3} M monovalent or >10^{-4} M divalent electrolyte).

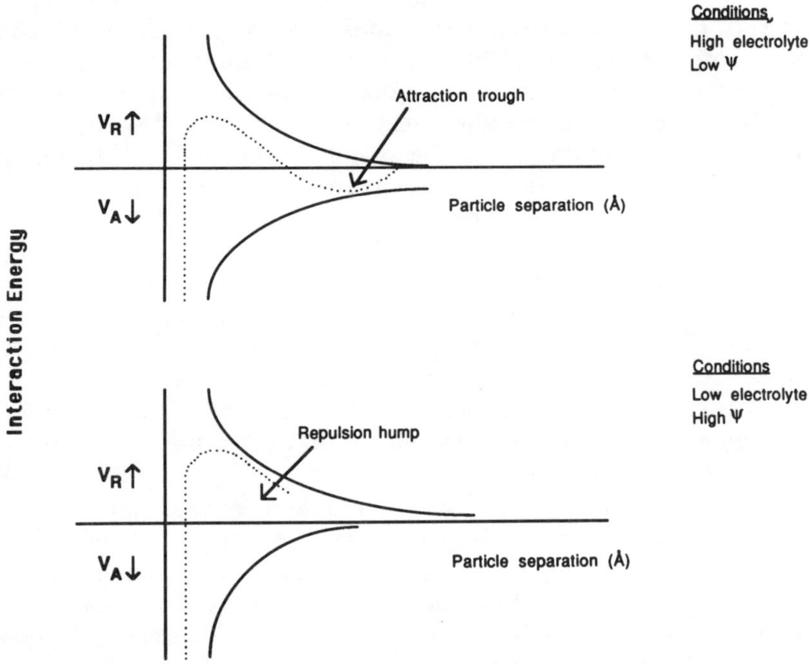

FIGURE 2. The interaction energy between a particle and substrate in electrolyte solutions.

In order for cells to become firmly (or irreversibly) attached to surfaces they must bridge the separation distance (usually on the order of 1 nm) depicted in Figure 3. One theory of bacterial adhesion to solid substrates states that irreversible adsorption occurs through the action of adhesive extracellular polysaccharides [23]. Others have shown that the strong interaction with hydrophobic adsorbents is probably mediated by cell-surface proteins; a different mechanism may exist for hydrophilic adsorbents [61,62].

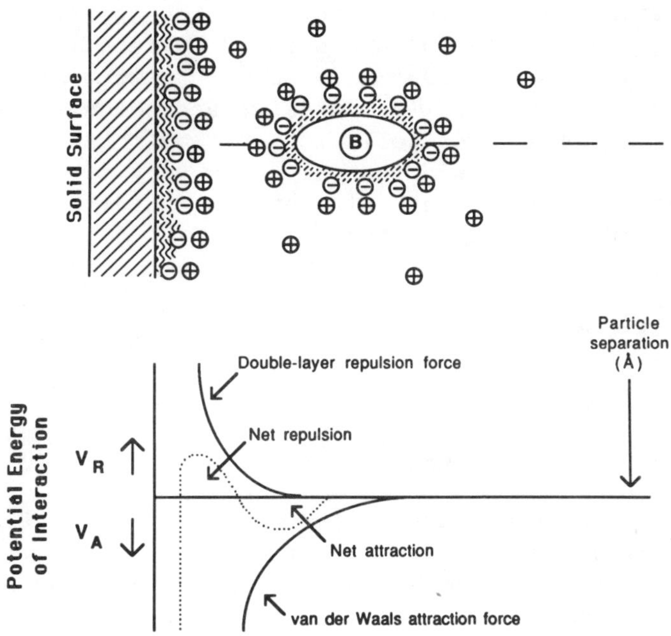

FIGURE 3. Reversible adsorption of a bacterium to a solid surface. The bacterium (B) is held, by electrostatic forces, a finite distance from the solid surface because of the net repulsive force of interaction (dotted line is the sum of V_R and V_A).

The surface physics of adhesion is difficult to model. Currently, the predicted strengths of adhesive joints are a great deal stronger than the actual adhesive strengths; there is apparently little correlation between the two [34]. Nevertheless, refinement of the models should improve predictions of adhesive strengths when better estimates of inherent flaws that exist in any adhesive joint are considered. These adhesive flaws are the primary source of weakness in macroscopic adhesive bonds. The adhesive joint between a bacterium and a solid surface is likely to be free of flaws because the bacterium itself is the source of the adhesive and because the area of adsorption (per cell) is very small [34]. The predicted strength of adhesion may very closely approximate the actual adhesive strength of the bond between bacterium and substrate.

The process of adhesion can be thermodynamically described in terms of surface free energy which is defined as the energy required to form a new unit area of surface:

$$\frac{dG}{dA} = \gamma$$

where G = Gibbs free energy, A = area and γ = surface free energy per unit volume [34]. An empirical way of visualizing γ is as follows. Consider an imaginary line, l, drawn through some liquid a distance, Δs, by a force, F. By definition,

$$\gamma = \frac{F\Delta s}{l\Delta s}$$

where $F\Delta s$ is a measure of energy in dynes x cm and $l\Delta s$ is the area (since $l\Delta s$ describes the area after the imaginary line, l, has been drawn through the liquid for the distance Δs). Thus, for liquids, the quantity γ is really force per unit length (F/l), which is called the coefficient of surface tension, and can be empirically described as the force required to part a unit length of surface.

The coefficient of surface tension for liquids can easily be measured. For example, a wettable solid (usually platinum) is drawn through the liquid-air interface and the force required to part a unit length of the interface is measured with a du Nory torsion balance. The coefficient of surface tension for liquids can also be determined by observing the rise or decline of the liquid in a capillary tube.

Unfortunately, the surface tension and the surface free energy of a solid are not necessarily equal. Furthermore, the types of tests used to determine the surface tension of liquids are obviously inappropriate for solids (e.g., the breaking strength of solids involves too many practical considerations to be accurate) [34]. The most successful method for measuring the surface free energy of solids is one described by Zisman [71,86,87]; an empirical parameter of wettability called the critical surface tension, γ_c, is determined by measuring the contact angles, θ, of a series of droplets, each with known and different surface tensions. The plot of cosine θ for each droplet versus the known surface tension of the droplet is fitted with the best straight line. The intercept of the line with the cosine $\theta = 1$ describes the value for critical surface tension of the substrate, because γ_c is the coefficient of surface tension under conditions in which the solid is completely wetted (Figure 4). The value, γ_c, is only an approximation of the solid-surface free energy. Usually, γ_c is somewhat less than the actual surface free energy of the solid surface because of error from a variety of interactions between the solid and droplets of the particular liquid used in the determination [34].

Young's equation describes the balance of forces that exist for a liquid drop on a surface [85]:

$$\gamma_{sv} = \gamma_{sl} + \gamma_{lv} \cos \theta$$

where the subscripts sv, sl and lv refer to the surface free energies between the solid-vapor, solid-liquid, and liquid-vapor interfaces, respectively. Good [35] showed experimentally that the interfacial tension between a liquid and a solid is described by:

$$\gamma_{sl} = \gamma_{sv} + \gamma_{lv} - 2\,\Phi_{sl}\,(\gamma_{sv}\,\gamma_{lv})^{1/2}$$

where Φ is the interaction parameter between the solid and the liquid, keeping in mind that the interfacial tension and free energy of liquids (γ_{lv}) are equivalent. The parameter, Φ_{sl}, is dependent upon the molecular properties of the two substances. If the intermolecular forces across the interface of the two phases are the same as they are within both the phases, then $\Phi = 1$. If the intermolecular forces are different between the phases, then $\Phi < 1$. If there is interaction between the phases (e.g., solubility), then $\Phi > 1$. Thus, the surface free energy of a solid surface, γ_{sv}, can be approximated from the wettability, γ_c, because $\gamma_{lv} = \gamma_c$ when cosine $\theta = 1$. The estimate of γ_{sv} is therefore dependent upon a good estimate of Φ_{sl}, which may be obtained from the value of dipole moment, ionization energies, and polarizability of both the liquid and the solid phases [34].

Investigators have reported a range of values for γ_c (20-30 dynes/cm) over which minimum bacterial adsorption occurs [3,30,31,36]. This "biocompatability range" of γ_c can be demonstrated <u>in situ</u> [30,31]. Biocompatable surfaces; i.e., surfaces which are less easily fouled, were colonized by only 1 to 10% as many bacteria as surfaces outside this γ_c range (Figure 2 in ref. 31). These observations were made by counting the attached bacteria (microscopic direct cell counting or biomass estimates) per unit area of solid surface after immersion in laboratory media or natural aquatic ecosystems. Recalling that, <u>in situ</u>, the first event in biofilm formation is the spontaneous adsorption of an organic film, we conclude from these experiments that γ_c must strongly influence the organic macromolecular conditioning films because it is to this conditioning film that bacteria actually adsorb.

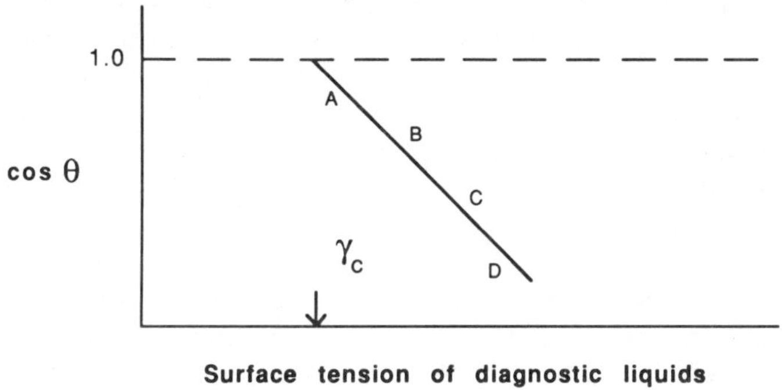

FIGURE 4. Critical surface tension, γ_c, is determined by contact angle measurements of a series of diagnostic liquids (e.g., A,B,C and D on platinum might be represented by n-hexane, 1-methylnaphtalene, α-bromonaphthalene, and methyle iodide, respectively).

A significant improvement in the determination of solid surface energies from contact angle measurements is based on an empirically-derived relationship between Φ and γ_{sl} [1,34]. Absolom _et al_. [1] were able to calculate all of the terms in a free energy balance of bacterial adhesion to solid subtrates. Their thermodynamic model allowed predictions of bacterial adhesion to various polymeric surfaces and of the surface tension of different bacterial genera.

Although no surfaces have been found that are impervious to biological fouling activities, surface structure does appear to influence the initial rate of fouling (i.e., over the first few hours of exposure).

In general, smoother surfaces foul at a slower initial rate than do rough surfaces. Once a primary biofilm is established, fouling proceeds at a rate that is largely independent of surface structure.

Fluid velocity appears to affect both the induction time and structure of fouling biofilms [54]. Increased fluid velocities retard the formation of biofilms. Filamentous bacteria tend to accumulate on test surfaces exposed to high fluid velocities. Filamentous forms appear after the deposition of large amounts of polymeric material on the surface. When the velocities and shear stress are low, filamentous forms appear, regardless of the presence of polymeric material.

Filamentous forms may develop from rod-shaped species such as pseudomonads, which elongate after a period of biofilm growth [78]. In some cases rod-shaped bacteria grow without dividing and develop into very long filaments up to 15-20 μm in length [53]. Filamentous forms, through their ability to penetrate the viscous sublayer, have a greater effect on the fluid frictional resistance than the biofilms composed of rod-shaped species [53]. Morphological changes in biofilms may, in part, be due to physiological adaptations of the adherent bacteria to environmental changes.

Biological Factors

Bacteria may overcome the electrostatic double-layer forces of repulsion that prevent contact with substrates by bridging the distance with polymers and/or appendages (Figure 3). It is this process that likely accounts for the tenacious adhesion of sessile bacteria. Biological factors associated with biofilm formation, therefore, contribute significantly to what often results in biofilm fouling. Recent advances in the field of microbial ultrastructure have demonstrated the exquisite form and function of the bacterial cell surface polymers and appendages that mediate biofilm formation.

The bacterial glycocalyx

Bacterial cell-surface structure has traditionally been studied using organisms from laboratory liquid cultures. In the absence of the selective pressures present in natural aquatic habitats, bacteria cell surfaces assume altered configurations. One example of such an alteration is the tendency to eliminate or reduce the production of an outer envelope component often refered to as glycocalyx. The glycocalyx forms the outermost layer of the cell envelope in most sessile bacteria observed in nature [16-25]. This structure also appears to contribute to the adhesion of the cells to surfaces [20-25]. Costerton _et al_. [23] applied the term "glycocalyx" to this bacterial envelope component as a result of the ultrastructural similarity it shared with the carbohydrate-rich "fuzzy coat" that surrounds certain animal cells [9]. Costerton later defined the bacterial glycocalyx as any polysaccharide-containing bacterial surface structure that is distal to the surface of the outer membrane of Gram-negative bacteria (the LPS-containing layer) or to the surface of the peptidoglycan layer of Gram-positive bacteria [24,25].

The glycocalyx of bacteria may be categorized by subdivision into two main types: the S-layer and the capsule. The S-layer glycocalyx has been observed in a few species (e.g., _Spirillum serpens_ and _Bacteroides ruminicola_) and has been described as a regular, two-dimensional array of globular glycoprotein units [72]. The majority of adherent bacterial species that have been observed in nature elaborate a polysaccharide matrix, commonly refered to as a capsule, which is up to 98% water [24,25,72]. Capsules have been further categorized by certain non-exclusive descriptors:

rigid, flexible, integral and peripheral. Rigid capsules are sufficiently organized so as to resist penetration of particulate dyes such as nigrosin or India ink; flexible capsules are not so endowed. Integral capsules are intimately associated with the cell surface such that centrifugation usually fails to disperse the exopolysaccharide; peripheral capsules are loosely bound and readily lost to the surrounding environment. This terminology provides a generally useful open-ended series of descriptors that is applicable for most of the known forms of bacterial exopolysaccharides. For example, the exopolysaccharide of Klebsiella aerogenes may be described as a rigid, integral glycocalyx whereas that of Pseudomonas aeruginosa is flexible and peripheral [24,25].

Some type of capsular glycocalyx can be observed around most of the cells in natural biofilms. Many are quite permeable to particulate dyes and, therefore, are not readily visible under the light microscope. Light-microscopy has been used, however, in concert with capsule-specific immunoglobulin proteins to render the fully hydrated flexible glycocalyx more refractile. This occurs as a result of an increase in density of the surface structure due to formation of an antibody-polysaccharide complex [7]. This antibody stabilization technique is known as the "Quellings reaction" [58] and was used as early as 1896 [68]. The elaborate organization and structure of capsules was not realized until the application of specific electron-dense stains in samples prepared for transmission electron microscopy (TEM). Stablization with capsule-specific immunoglobulins followed by treatment with ruthenium red (an electron-dense, cationic stain) has proven to be a useful technique to preserve and visualize anionic capsules in bacteria [20].

A glycocalyx provides a microorganism with various environmental advantages beyond its role as a holdfast. The structure promotes survival of the organism in low nutrient environments and resistance to bacteriophage attack and the action of biocides [21]. The structure and function of the bacterial glycocalyx has been recently reviewed by Costerton *et al*. [24,25]. The capsules of Gram-positive bacteria such as *Bacillus subtilis* are comprised of teichoic acids. Teichoic acids consist of a medley of subunits, including polyol phosphates (glyerol phosphate or ribitol phosphate) and carbohydrates such as glucose, galactose, N-acetylglucosamine and N-acetylgalactosamine [10]. The capsules of Gram-negative bacteria may be composed of a wide variety of homo- and heteropolysaccharides [73-75,82]. Some strains produce several polymers simultaneously [25]. Although capsules are common features of biofilm-forming bacteria, there is experimental evidence which suggests that they are not involved in the initial events of irreversible adsorption of the cells to the surface [61,62].

Fimbriae

In 1950, Houwink and van Iterson [41] described surface structures on *E. coli* that are morphologically distinct from flagella; the suggestion that these structures act as adhesins dates to the original observation. Later, the structures were renamed fimbriae (= fibers) and several types were described according to their morphological characteristics. The most commonly observed fimbria was designated type 1. This designation is still used today. The name pili (= hairs) was suggested by Brinton in 1959 [11] to describe several other surface structures, some of which are susceptible to bacteriophage attack. The distinction between fimbriae and pili has been blurred since their first description. The terms are generally considered synonomous but a distinction according to function is now being adopted [60]. Accordingly, there are presently two recognized groups of non-flagellar surface structures on E. coli: Group 1 consists of the adhesive factors (fimbriae); Group 2 structures include all of the surface organelles that mediate conjugal chromosomal transfer (i.e., the sex pili) and are typed according to phage susceptibility. Generally, E. coli has only one or two sex pili per cell in contrast to 100-400 fimbriae per cell. Fimbriae or pili are rigid, needle-like structures that mediate attachment of some bacteria to tissues. The morphology of these appendages is ideally suited to overcome the energy barrier that exists at a surface. In spite of this, fimbriae are not likely to be important in adhesion of industrial biofilms and have probably been mistakenly identified in scanning electron micrographs (SEM) of biofilm bacteria due to insufficient resolution.

Other cell surface structures involved in adhesion

Flagellar adsorption to glass slides may mediate bacterial attachment under certain conditions [55]. Transmission electron micrographs published by Weiss et al. [81] demonstrated that bacteria of the genus Sulfolobus which are chemolithotrophic, acidophilic, thermophilic, Gram-negative, lobed cells that use elemental sulfur as an energy source, adhere to sulfur crystals in hot springs by non-flagellar appendages. The sulfur crystals to which Sulfolobus adhere become the substrate for metabolism, in which case, the sulfur is most certainly not an inert adsorbent.

Goblet-shaped proteinaceous subunits from the cell wall of the gliding bacterium, Flexibacter polymorphus, probably act as cell surface "pores" through which some type of adhesin is excreted, (perhaps a glycoprotein, H.F. Ridgway, personal communication) thus mediating gliding motility on surfaces [65-67]. The adhesin is extruded through a duct located in the stem of the goblet-shaped subunit; both of these structures are associated with the outer lipopolysaccharide component of the cell membrane.

Bacteria in certain genera are able to adhere to solid surfaces by means of stalks or holdfast surface processes. The genus *Caulobacter* is commonly found on submerged solid substrates, particularly in low nutrient, freshwater environments [63]. In fact, an easy way to observe *Caulobacter* in freshwater habitats is simply to suspend glass microscope slides in the water for several days and observe the adherent biofilm microscopically: *Caulobacter* often adheres to surfaces in regular groups called rosettes.

Apparently, a polysaccharide adhesin is secreted from the distal end of the stalk. The cells in rosettes are separated by a common mass of this adhesin which is located at the center of the rosette [69]. Bacteria with complex cell cycles and surface structures such as *Caulobacter* and *Hyphomicrobium* may require rather complex growth factors [19]. Hence, Corpe [17,19] reasoned that the prosthecate bacteria in biofilms are probably not the initial or primary colonizers of the surface, but rather, they are "secondary periphytes". That is, they are genera that adhere after a primary biofilm has been established. There is experimental evidence to support this hypothesis [17,52]. Thus, the development of biofilms in nature involves a succession of bacterial genera on surfaces exposed to both flowing and quiescent water.

Metabolism in bacterial biofilms

Bacteria in biofilms are generally more metabolically active than their planktonic (free-floating) counterparts. Ladd *et al*. [48] showed that biofilm bacteria in freshwater streams are about 7.9 times more active than the planktonic bacteria in the same stream; there is considerable evidence to support this conclusion [40,43-45,64,70]. Studies that utilized in situ sediment traps in seawater showed that bacteria associated with large sedimenting particles (> 35 μm) may contribute as much as 100 times more biomass carbon than planktonic bacteria [32]. Many researchers have demonstrated a strong correlation between heterotrophy and attached bacteria in freshwater [8]. Aside from these limited studies in natural waters, the physiology of adherent microorganisms in biofilms is largely unknown. This has made it difficult to predict the conditions that stimulate or inhibit biofilm formation in industrial water systems. In addition, industrial water systems are largely inaccessible to outside scientists that may have an interest in this area of microbiology. Collaborative efforts by in-house personnel and outside specialists are needed to advance our knowledge in this field.

References

1. Absolom, D.R., F.V. Lamberti, Z. Policova, W. Zingg, C.J. van Oss, and A.W. Neumann. 1983. Surface thermodynamics of bacterial adhesion. Appl. Environ. Microbiol. 46:90-97.

2. Bader, R.G., D.W. Hood, and J.B. Smith. 1960. Recovery of dissolved organic matter in seawater and organic sorption by particulate material. Geochim. Cosmochim. Acta. 19:236-243.

3. Baier, R.E. 1980. Substrata influences on adhesion of microorganisms and their resultant new surface properties. p. 59-104. In G. Bitton and K. C. Marshall (ed.), Adsorption of microorganisms to surfaces. John Wiley & Sons, New York.

4. Baier, R.E. 1982. Cell adhesion to biomaterials

surfaces: conflicts and concerns. J. Biomed. Mater. Res. 16:173-175.

5. Baier, R.E. and G.I. Loeb. 1971. Multiple parameters characterizing interfacial films of a protein analogue. p. 79-96. In C.D. Craven (ed.), Polymer characterization: interdisciplinary approaches. Plenum, New York.

6. Baier, R.E. and L. Weiss. 1975. Demonstration of the involvement of adsorbed proteins in cell adhesion and cell growth on solid surfaces. Adv. Chem. Series. 145:300-307.

7. Bayer, M.E. and H. Thurow. 1977. Polysaccharide capsule of *Escherichia coli*: Microscope study of its size, structure and sites of synthesis. J. Bact. 130:911-936.

8. Bell, C.R. and L.J. Albright. 1981. Attached and free-floating bacteria in the Fraser River Estuary, British Columbia, Canada. Mar. Ecol. Prog. Ser. 6:317-327.

9. Bennett, H.S. 1963. Morphological aspects of extracellular polysaccharides. J. Histochem. Cytochem. 11:14-25.

10. Birdsell, D.C., R.J. Doyle and M. Morgenstern. 1975. Organization of teichoic acid in the cell wall of *Bacillus subtilis*. J. Bact. 121:726-734.

11. Brinton, C.C. 1959. Non-flagellar appendages of bacteria. Nature 183:782-786.

12. Characklis, W.G. 1973. Attached microbial growths-I. Attachment and growth. Water Res. 7:1113-1127.

13. Characklis, W.G. 1981. Fouling biofilm development: a process analysis. Biotechnol. Bioengin. 23:1923-1960.

14. Characklis, W.G., and K.E. Cooksey. 1983. Biofilms and microbial fouling. Adv. Appl. Microbiol. 29:93-138.

15. Chet, I., P. Asketh, and R. Mitchell. 1975. Repulsion of bacteria from marine surfaces. Appl. Environ. Microbiol. 15:1043-1045.

16. Corpe, W.A. 1970. Attachment of marine bacteria to solid surfaces. p. 73-87. In R.S. Manly (ed.), Adhesion in biological systems. Academic Press, New York.

17. Corpe, W.A. 1972. Microfouling: The role of primary film forming marine bacteria. p. 598-605. In Proceedings of the 3rd international congress on marine corrosion and fouling, National Bureau of Standards, Gaithersburg, MD.

18. Corpe, W. A. 1974. Detachment of marine periphytic bacteria from surfaces of glass slides. Dev. Ind. Microbiol. 15:281-287.

19. Corpe, W. A. 1980. Microbial surface components involved in adsorption of microorganisms onto surfaces, p. 105-144. In G. Bitton and K.C. Marshall (ed.), Adsorption of microorganisms to surfaces. John Wiley & Sons, New York.

20. Costerton, J.W. 1980. Some techniques involved in study of adsorption of microorganisms to surfaces, p. 429-430. In G. Britton and K. C. Marshall (ed), Adsorption of microorganisms to surfaces. John Wiley & Sons, New York.

21. Costerton, J.W. and K.-J. Cheng. 1975. The role of the bacterial cell envelope in antibiotic resistance. J. Antimicrob. Chemother. 1:363-377.

22. Costerton, J.W. and G.G. Geesey. 1979. Microbial contamination of surfaces, p. 211-221. In K.L. Mittal (ed.), Surface contamination, Vol. I. Plenum Publishing Corp., New York.

23. Costerton, J.W., G.G. Geesey and K.-J. Cheng. 1978. How bacteria stick. Sci. Am. 238:86-95.

24. Costerton, J.W. and R.T. Irvin. 1981. The bacterial glycocalyx in nature and disease. Ann. Rev. Microbiol. 35:299-324.

25. Costerton, J.W., R.T. Irvin and K.-J. Cheng. 1981. The role of bacterial surface structures in pathogenesis. CRC Crit. Rev. Microbiol. 8:303-338.

26. Curtis, A.S.G. 1960. Cell contacts: some physical considerations. Amer. Nat. 94:37-56.

27. Curtis, A.S.G. 1962. Cell contact and adhesion. Biol. Rev. 37:82-129.

28. Daniels, S.L. 1980. Mechanisms involved in sorption of microorganisms to solid surfaces, p. 7-58. In G. Bitton and K.C. Marshall (ed.), Adsorption of microorganisms to surfaces. John Wiley & Sons, Inc., New York.

29. Derjaguin, B.V. and L. Landau. 1941. Theory of the stability of strongly charged lyophobic sols and of the adhesion of strongly charged particles in solutions of electrolytes. Acta Physicochem. U.R.S.S. 14:633.

30. Dexter, S.C. 1979. Influence of substratum critical surface tension on bacterial adhesion - in situ studies. J. Colloid and Interface Sci. 70:346-354.

31. Dexter, S.C., J.D. Sullivan, J. Williams and S.W. Watson. 1975. Influence of substrate wettability on the attachment of marine bacteria to various surfaces. Appl. Environ. Microbiol. 30:298-308.

32. Fellows, D.A., D.M. Karl and G.A. Knauer. 1981. Large particle fluxes and the vertical transport of living carbon in the upper 1500m of the northeast Pacific Ocean. Deep-Sea Res. 28A:921-936.

33. Flaig, W. 1971. Some physical and chemical properties of humic substances as a basis of their characterization. Adv. Org. Geochem. 33:49-67.

34. Gerson, D.F. and J.E. Zajic. 1979. The biophysics of cellular adhesion. p. 29-57. <u>In</u> K. Venkatsubramanian (ed.), Immobilized microbial cells. American Chemical Society symposium series, no. 106, Am. Chem. Soc., Washington, D.C.

35. Good, R.J. 1964. Theory for the estimation of surface and interfacial energies VI. Surface energies of some flurocarbon surfaces from contact angle measurements. Adv. Chem. 43:74-87.

36. Grinnel, F. 1978. Cellular adhesiveness and extracellular substrata. Int. Rev. Cytol. 53:65-144.

37. Harvey, H.W. 1941. On changes taking place in sea water during storage. J. Mar. Biol. Assoc. U.K. 25:225-233.

38. Haydon, D.A. 1961. The surface charge of cells and some other small particles as indicated by electrophoresis. I. The zeta potential-surface charge relationships. Biochim. Biophys. Acta. 50:450-457.

39. Haydon, D.A. 1961. The surface charge of cells and some other small particles as indicated by electrophoresis. II. The interpretation of the electrophoretic charge. Biochim. Biophys. Acta. 50:457-462.

40. Hodson, R.E., A.E. Maccubbin and L.R. Pomeroy. 1981. Dissolved adenosine triphosphate utilization by free-living and attached bacterioplankton. Mar. Biol. 64:43-51.

41. Houwink, A.L. and W. van Iterson. 1950. Electron microscopical observations on bacterial cytology. Biochim. Biophys. Acta. 5:10-44.

42. Hunt, J.R. 1982. Particle dynamics in seawater: Implications for predicting the fate of discharged particles. Environ. Sci. Technol. 16:303-309.

43. Jannasch, H.W. 1967. Growth of marine bacteria at limiting concentrations of organic carbon in seawater. Limnol. Oceanogr. 12:264-271.

44. Jeffrey, W.H. and J.H. Paul. 1986. Activity of an attached and free-living *Vibrio* sp. as measured by thymidine incorporation, p-iodonitrotetrazolium reduction and ATP/DNA ratios. Appl. Environ. Microbiol. 51:150-156.

45. Jeffrey, W.H. and J.H. Paul. 1986. Activity measurements of planktonic microbial and microfouling communities in a eutrophic estuary. Appl. Environ. Microbiol. 51:157-162.

46. Keller, E.F. and J. Segel. 1971. Traveling bands of chemotactic bacteria: A theoretical analysis. J. Theor. Biol. 30:235-248.

47. Kim, S W. and R.G. Lee. 1975. Adsorption of blood proteins onto polymer surfaces. Adv. Chem. Ser. 145:218-229.

48. Ladd, T.I., J.W. Costerton and G.G. Geesey. 1979. Determination of the heterotrophic activity of epilithic microbial populations. p. 180-195. In J.W. Costerton and R.R. Colwell (ed.), Native Aquatic Bacteria. ASTM STP 695, Philadelphia, PA.

49. Lapidus, I.R. 1980. Microbial chemotaxis in flowing water in the vicinity of a source of attractant or repellent. J. Theor. Biol. 85:543-547.

50. Loeb, G.I. and R.A. Neihof. 1975. Marine conditioning films. Adv. Chem. Ser. 145:319-335.

51. Marshall, K.C., R. Stout and R. Mitchell. 1971. Mechanism of the initial events in the sorption of marine bacteria to surfaces. J. Gen. Microbiol. 68:337-348.

52. Marszalek, D.S., S.M. Gerchakov and L.R. Udey. 1979. Influence of substrate composition on marine microfouling. Appl. Environ. Microbiol. 38:987-995.

53. McCoy, W.F., J.D. Bryers, J. Robbins, and J.W. Costerton. 1981. Observations of fouling biofilm formation. Can. J. Microbiol. 27:910-917.

54. McCoy, W.F. and J.W. Costerton. 1982. Fouling biofilm development in tubular flow systems. Dev. Ind. Microbiol. 23:551-558.

55. Meadows, P.S. 1971. The attachment of bacteria to solid surfaces. Arch. Microbiol. 75:274-281.

56. Neihof, R.A. and G.I. Loeb. 1972. The surface charge of particulate matter in seawater. Limnol. Oceanogr. 17:7-16.

57. Neihof, R.A. and G.I. Loeb. 1974. Dissolved organic matter in seawater and the electric charge of immersed surfaces. J. Mar. Res. 32:5-12.

58. Neufeld, F. 1902. Uber die agglutination der Pneumokokken und uber die theorieen der agglutination. Z. fur Infektionskrankheiten. Parasitare Krankheiten und Hygiene der Haustiere. 40:54-72. (ref. in Boyer J. Bacteriol. 130:911-936.)

59. O'Melia, C.R. 1980. Aquasols: The behavior of small particles in aquatic systems. Environ. Sci. Technol. 14:1052-1060.

60. Orskov, I., F. Orskov, and A. Birch-Andersen. 1980. Comparison of _E. coli_ fimbrial antigen F7 with type 1 fimbriae. Inf. Immun. 27:657-666.

61. Paul, J.H., and W.H. Jeffrey. 1985. Evidence for separate adhesion mechanisms for hydrophilic and hydrophobic surfaces in _Vibrio proteolytica_. Appl. Environ. Microbiol. 50:431-437.

62. Paul, J.H., and W.H. Jeffrey. 1985. The effect of surfactants on the attachment of estuarine and marine bacteria to surfaces. Can. J. Microbiol. 31:224-228.

63. Poindexter, J.S. 1981. The caulobacters: ubiquitous unusual bacteria. Microbiol. Rev. 45:123-179.

64. Pomeroy, L.R. and R.E. Johannes. 1968. Occurrence and respiration of ultraplankton in the upper 500 meters of the ocean. Deep-Sea Res. 15:381-391.

65. Ridgway, H.F. and R.A. Lewin. 1973. Goblet-shaped subunits from the wall of a marine gliding microbe. J. Gen. Microbiol. 79:119-128.

66. Ridgway, H.F. and R.A. Lewin. 1983. Subunit composition of goblet-shaped particles from the cell wall of *Flexibacter polymorphus*. Can J. Microbiol. 29:1689-93.

67. Ridgway, H.F., R.M. Wagner, W.T. Dawsey and R.A. Lewin. 1975. Fine structure of the cell envelope layers of *Flexibacter polymorphus*. Can. J. Microbiol. 21:1733-1750.

68. Roger, D.H. 1896. Les infections non bacteriennes. Rev. Gen. Sci. Appl. 7:770-775.

69. Rutter, P.R. 1980. The physical chemistry of bacterial adhesion. p. 103-135. *In* A.S.G. Curtis and J.D. Pitt (ed.), Cell adhesion and motility, Cambridge University Press, New York.

70. Seki, H. 1972. The role of microorganisms in the marine food chain with reference to organic

aggregates. Mem. Ist. Ital. Idrobiol. 29 (suppl):245-25.

71. Shafrin, E.G. and W.A. Zisman. 1967. Effect of adsorbed water on the spreading of organic liquids on soda-lime glass. J. Am. Ceramic Soc. 50:478-484.

72. Sleytr, U.B. 1978. Regular array of macromolecules on bacterial cell walls: structure, chemistry, assembly, and function. Int. Rev. Cytol. 53:1-64.

73. Sutherland, I.W. 1977. Bacterial exopolysaccharides-their nature and production, p. 25. In I.W. Sutherland (ed.), Surface carbohydrates of the prokaryotic cell. Academic Press, London.

74. Sutherland, I.W. 1980. Polysaccharides in the adhesion of marine and freshwater bacteria, p. 329-338. In R.C.W. Berkeley, J.M. Lynch, J. Melling, P.R. Rutter, and B. Vincent (ed.), Microbial adhesion to surfaces. Ellis Horwood Publishers, Chichester, U.K.

75. Sutherland, I.W. 1983. Microbial exopolysaccharides-their role in microbial adhesion in aqueous systems. Crit. Rev. Microbiol. 10:173-201.

76. Verwey, E.J.W. and J.T.G. Overbeck. 1948. Theory of the stability of lyophobic colloids. Elsevier, Amsterdam.

77. Ward, J.B. and R.C.W. Berkeley. 1980. The microbial cell surface and adhesion. p. 47-66. In R.C.W. Berkeley, J.M. Lynch, J. Melling, P.R. Rutter, and B. Vincent (ed.), Microbial adhesion to surfaces. Soc. Chem. Ind., Ellis Horwood

Ltd., Chichester, U.K.

78. Wardell, J.N., C.M. Brown and D.C. Ellwood. 1980. A continuous culture study of the attachment of bacteria to surfaces. p. 221-229. In R.C.W. Berkeley, J.M. Lynch, J. Melling, R.R. Rutter, and B. Vincent (ed.), Microbial adhesion to surfaces, Ellis Horwood Ltd, Chichester, U.K.

79. Weiss, L. 1968. Studies on cellular adhesion in tissue culture. IX. Electrophoretic mobility and contact phenomena. Exp. Cell Res. 51:609-625.

80. Weiss, L. 1968. Studies on cellular adhesion in tissue culture. X. An experimental and theoretical approach to interaction forces between cells and glass. Exp. Cell Res. 53:603-614.

81. Weiss, R.L. 1973. Attachment of bacteria to sulphur in extreme environments. J. Gen. Microbiol. 77:501-507.

82. Wilkinson, J.F. 1958. The extracellular polysaccharides of bacteria. Bact. Rev. 22:46-73.

83. Young, L.Y. and R. Mitchell. 1972. The role of chemotactic responses in primary microbial film formation. p. 617-623. In Proceedings of the 3rd international congress on marine corrosion and fouling, National Bureau of Standards, Gaithersburg, MD

84. Young, L.Y. and R. Mitchell. 1973. Negative chemotaxis of marine bacteria to toxic chemicals. Appl. Microbiol. 25:972-975.

85. Young, T. 1805. An essay on the cohesion of fluids. Phil. Trans. Roy. Soc. 95:65-87.

86. Zisman, W.A. 1964. Relation of the equilibrium contact angle to liquid and solid constitution. Adv. Chem. 43:1-51.

87. Zisman, W.A. 1972. Surface energetics of wetting, spreading, and adhesion. J. Paint Tech. 44:41-57.

88. Zobell, C.E. 1943. The effect of solid surfaces upon bacterial activity. J. Bacteriol. 46:39-59.

CHAPTER 3. MICROBIALLY INFLUENCED CORROSION

J.W. Costerton and Joe Boivin, Microbios Ltd., Calgary, Alberta, Canada

The refining of metal ores and the manufacture of usable alloys involves the input of energy. The natural tendency for the elements in manufactured goods to give up this energy and revert to a more stable state is the process of corrosion.

The corrosion of metals is an electrochemical phenomenon. Corrosion involves the interaction of the solvent (usually water) with anodic sites (where metals give up electrons and go into solution) and cathodic sites (where the electrons produced at the anode are consumed). The metal itself provides a path for the electron flow under the influence of the potential generated between the anode and cathode (Figure 1).

The kinetics of the reaction are affected by the constitution of the bulk solution in general corrosion and local effects at the anode and cathode in pitting corrosion. A metal surface in contact with water will have a multitude of micro areas acting as anodes and cathodes. These areas shift in location as the reaction products accumulate at the anode and cathode, reducing the effective potential. In the absence of local effects, this shifting of corroding areas results in a generalized corrosion loss over the entire surface of the metal.

This general metal loss is often tolerable since substantial attack can take place before the structural integrity of the material is affected. When this attack is restricted to a small area, however, rapid loss of metal can result in early failure of a component.

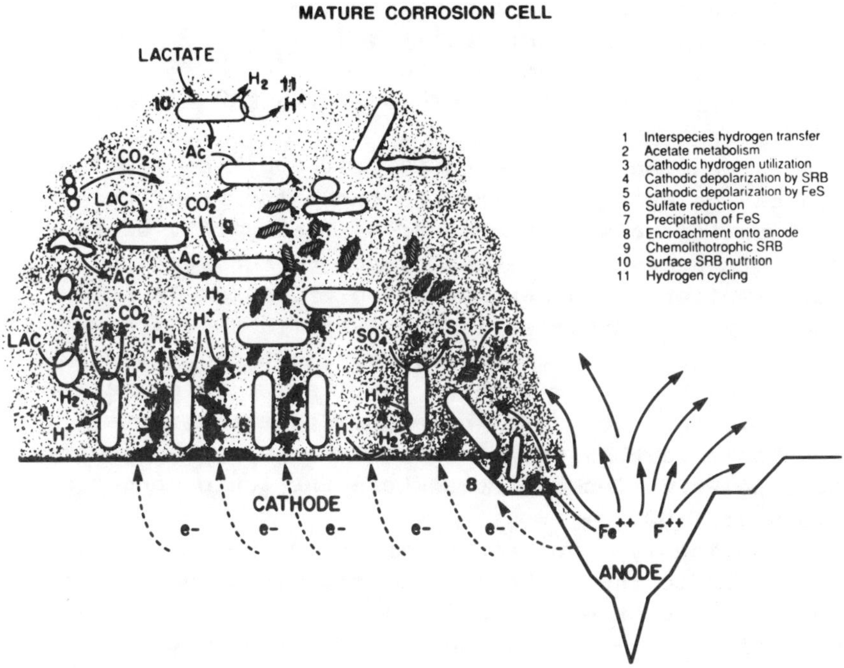

FIGURE 1. Schematic representation of microbial influenced corrosion.

Mechanism of Localized Corrosion

Cathodic Reactions

The electrons produced by anodic reaction ($M^o \rightarrow M^{n+} + ne^-$, where M is any metal) are consumed at the cathode by several different reactions dependent on the nature of the environment. These are:

Oxygen Reduction: $O_2 + 4H^+ + 4e^- \rightarrow 2H_2O$
(acid solutions)
Oxygen Reduction: $O_2 + 2H_2O + 4e^- \rightarrow 4OH^-$
(neutral or alkaline solutions)
Proton reduction: $2H^+ + 2e^- \rightarrow H_2$
Metal ion reduction: $Fe^{+3} + e^- \rightarrow Fe^{+2}$
Metal Deposition: $Cu^{+2} + 2e^- \rightarrow Cu^o$

These reactions will be rate limited by the availability of the energy required to cause the reaction (activation polarization) and by the ability of reactants to access the cathodic area (concentration polarization). Concentration polarization is influenced by the number of ions present in solution or by accumulation of reaction products that block the path to the cathode. In general, activation polarization controls reaction rate where the concentration of reactants is high (e.g. acid solutions) and concentration polarization controls where the ionic concentrations are low.

Obviously, the rate at which the anode provides electrons to the cathode also determines the rate of reaction and anodic polarization reactions exist which will affect the rate of corrosion. Most microbial corrosion reactions exploit the cathodic processes since this is the area where the energy of the corrosion process is discharged.

Localization of attack

The above reactions can take place over large areas of metal surface. Localization of attack occurs when environmental effects induce heterogeneities on the metal surface. These can be:

1) Differences in pH or salt concentration: the area of reduced pH or elevated salt concentration is attacked.

2) Differences in oxygen concentration: the area in contact with lower oxygen concentration is attacked.

3) Differences in concentration of other cathodic reductants: the area in contact with lower concentrations is corroded.

4) Local increases in cathodic reaction rates: the area of increased reaction is protected and adjacent areas are corroded.

Mechanical and metallurgical conditions on the surface of the metal will cause initiation of localized attack where a small area becomes more reactive due to the chemistry of the metal (e.g., sulfide inclusions) or an input of energy (e.g., cold working or thermal stresses) [3].

Rate of Attack

Corrosion rates are dependent upon the rate of electron production and consumption. These factors are regulated by anodic/cathodic depolarization and surface area. As the rate of production of electrons at the anode increases, metal loss is facilitated. Cathodic depolarization, resulting in consumption of electrons, is a significant factor in microbially influenced corrosion activity.

If the anodic area is large and the cathodic area is small, the attack will be generalized over a large area. Concentration polarization will reduce the rate of electron consumption and corrosion rates will be limited. Conversely, if the cathodic area is large and the anodic area small, corrosion will be localized. The rate of metal loss will be high in the anodic area (Figure 1).

Bacterial Evolution and MIC

Bacterial evolution reflects the competition for available energy sources. Those organisms best adapted for a particular niche proliferated because they outcompeted other organisms. As life forms became more complex, the diversity of species increased to reflect the more varied energy sources available. The evolution of photosynthesis altered the environment by the production of oxygen, allowing more efficient utilization of energy by those organisms able to utilize oxygen as a terminal electron acceptor. Those organisms not able to function in the presence of oxygen were restricted to niches where oxygen was depleted (e.g., sediments) and adapted to metabolize those substrates available to them.

Bacteria are ubiquitous in our environment; they are capable of utilizing an amazing array of energy sources and tolerate extreme conditions of temperature, pressure and chemical composition. This adaptability has profound implications in the processes of microbially influenced corrosion. The metallic structures developed by mankind, and the environments they contain, have provided unique opportunities to those organisms pre-adapted by nature to exploit them.

Microorganisms can influence corrosion in many ways:

1) Utilization of oxygen: When aerobic organisms colonize a surface they consume oxygen, preventing its access to the metal surface. As we have seen, oxygen-deficient areas are anodic to the metal and corrosion will result.

2) Utilization of hydrogen: Certain organisms, most notably the sulfate-reducing bacteria, can utilize cathodic hydrogen. The resulting depolarization of the cathode will increase the rate of metal loss at the anode (Figure 1).

3) Utilization of protective coatings: The ability of bacteria to degrade complex organic materials can result in destruction of protective coatings placed on metal surfaces. Where coatings are removed in patches, the area exposed to the corrosive environment becomes a small anode connected to a large cathode (the remaining coated area) and rapid metal loss ensues.

4) Utilization of inhibitors: Chemicals added to a system to protect the metal (e.g., corrosion inhibitors) or regulate anode/cathode reactions (e.g., nitrates) can be metabolized by certain bacteria resulting in loss of inhibition [9].

5) Production of corrosive metabolites: The end-product of metabolism of some microorganisms are often organic acids (e.g., acetic acids) or inorganic acids (e.g., H_2SO_4). When these products are concentrated in one area, rapid local metal dissolution can occur [9].

6) Production of concentration cells: Growth of microorganisms on surfaces can result in concentration of specific ions. These localized differences in concentration can shift the potential of metal surfaces resulting in creation of localized corrosion cells [19].

7) Removal of metal ions: Metabolic by-products (such as H_2S) can precipitate metal ions; this will depolarize the anode and allow the corrosion process to progress. Similarly, metal ions

can be tied up by the bacterial polymers, effectively removing them from the local environment.

Biofilms and Corrosion

Corrosion is a surface phenomenon. The development of the biofilm theory has allowed a fuller understanding of the relationship between microorganisms and the corrosion processes. Biofilms have several properties that influence localized corrosion on metal surfaces:

1) The establishment of aerobic colonies allows the creation of permanent sites of oxygen depletion on metal surfaces. Adjacent sites where oxygen can access the metal surface allow electron flow to proceed. Pitting of the area beneath the aerobic colonies ensues.

2) The scavenging of oxygen by the top layers in a biofilm community allows anaerobic organisms to develop at the metal surface. Many types of anaerobes can further exacerbate the corrosion process by production of corrosive metabolites or by depolarizing anodes and cathodes. The metabolic byproducts of the aerobes can also act as substrates for other organisms.

3) The biopolymers can create concentration cells on surfaces by trapping ions or otherwise modifying local metal surfaces. These differences in surface composition generate corrosion potentials between adjacent sites [19].

4) The electrical nature of the biofilm likely assists in the flow of electrons and cathodic reactants at metal surfaces. Leakage of enzymes into the biofilm and trapping of conductive

reaction products into a structured, hydrated structure most likely facilitates these processes.

5) The protective nature of the glycocalyx insulates the microorganism from deleterious substances naturally present in the environment or added by man to control them. This allows the corrosion process to continue unabated [5].

Organisms Involved in MIC

Eukaryotes

Macro-organisms: Fouling of metal surfaces, particularly in seawater, allows the establishment of sessile organisms such as barnacles, worms, molluscs, etc. on surfaces. The areas beneath their points of attachment become oxygen depleted and pitting results [16].

Fungi: The fungi have been implicated in the corrosion of fuel tanks. These organisms are capable of utilizing the hydrocarbons in diesel and jet fuels producing organic acids. They are capable of living in the small amounts of water present as a contaminant from carryover and condensation. The organism most often cited is *Cladosporium resinae* which caused severe structural damage to integral fuel tanks in jet aircraft in the 1950's and 1960's [13]. The problem has been controlled by the inclusion of large concentrations of biocide and meticulous housekeeping to prevent the influx of water.

The fungi have also been implicated in the destruction of natural and synthetic polymers applied as protective coatings for metallic surfaces. The deterioration of wooden members in cooling towers is also aggravated by fungi [13].

Procaryotes

Slime forming bacteria: Most bacteria produce exopolymers and attach to surfaces. The characteristics of the particular polysaccharide will determine the nature of the MIC associated with the microorganism. A typical and frequently cited genus of slime forming bacteria is _Pseudomonas_. These organisms can utilize an extremely wide range of substrates from simple organics to complex heterocyclic hydrocarbons and most importantly many of the currently used biocides. An analytical method has been developed using _Pseudomonas putida_ to allow culturing of anaerobic bacteria in the presence of biocide containing waters; the _P. putida_ scavenges oxygen from the culture and destroys any residual biocide [2]!

Sulfur bacteria: These organisms utilize sulfides and sulfur in the presence of oxygen to produce sulfuric acid. They are a major environmental pollutant in abandoned mining areas where they cause acidification of surface waters and liberation of toxic metals by dissolution of ores. This process has been harnessed and a large percentage of copper production is the result of microbial leaching of low-grade ores [1]. The production of sulfuric acid by these bacteria can result in extensive degradation of stone, concrete and steel structures.

Iron fixing bacteria: These bacteria use ferrous ion as an energy source, converting it to ferric hydroxide. They produce voluminous sheaths of iron products which adhere tightly to metal surfaces creating large tubercles. The area beneath these tubercles becomes depleted in oxygen resulting in pitting beneath the deposits. These tubercles often become colonized by anaerobic organisms, resulting in aggravated corrosion [6]. The tubercles are extremely protective and frequently harbor pathogens in municipal water mains [8]. The build-up of tubercles reduces the flow efficiency in pipelines--pigging is often required to remove them.

The iron fixing bacteria are frequently associated with the microbial corrosion of stainless steels. The anaerobic area beneath the tubercle is highly anodic in stainless steels which require the presence of oxygen to maintain proximity. The biofilm also concentrates chlorides, which further aggravate stainless steel corrosion activities. Rapid and extensive failures of stainless steel vessel and exchanger piping are frequently ascribed to the action of these bacteria, often because untreated water has been used for hydrostatic testing [18].

Sulfate reducing bacteria: The sulfate reducing bacteria (SRB) provide a classic example of the adaptation of a bacterium to a man-made environment that provides all of the growth factors needed by the organism. SRB are the major bacteria involved in the final mineralization of organic compounds in the environment. Recent investigations have shown them to be the dominant organism in freshwater and marine sediments and a very important part of the sulfur cycle in the ecosystem [7].

The SRB are an ancient organism. Their action is detectable in some of the oldest rocks on our planet and a sulfate reducing archaebacterium has recently been isolated. They are the most prevalent corrosion causing organism cited in MIC and have been implicated in so many instances of MIC that their presence is often considered a marker for biological corrosion just as *Escherichia coli* is used as an indicator of fecal contamination.

The prevalence of SRB in microbiological corrosion can be explained by their position in the ecosystem. The SRB are found in anaerobic sediments where they metabolize a wide range of substrates using sulfate as a terminal electron acceptor and produce

H_2S [13]. The key to their prevalence in this environment is the additional ability to utilize the hydrogen produced by commensal organisms, or within their own cells, to allow them to metabolize otherwise inaccessible substrates. They outcompete other organisms such as methanogens which utilize similar energy sources because they have a greater affinity for hydrogen [14]. The SRB can utilize this hydrogen as an additional energy source or to assist in the metabolism of otherwise unavailable nutrients such as acetate [12]. The SRB produce an enzyme called hydrogenase; this enzyme is present in the cytoplasm of the cell and also in the cell membrane. The presence of cathodic hydrogen thus allows the SRB to outcompete other organisms in the corroding area.

Let us consider the process of microbial corrosion in an oilfield water. Conditions are ideal for the SRB: the system is anaerobic, sufficient sulfate is present, and organic carbon sources are present in large amounts. Acetate is also generally present in substantial amounts, commonly up to several hundred ppm and sometimes over 1000 ppm. The environment closely parallels that of the SRB's normal ecological niche - marine sediments.

In such a system, a biofilm will develop rapidly. Absorption of organic and inorganic nutrients will allow growth of many species, but those best adapted to conditions in a micro-area will eventually dominate. This micro-environment will vary due to flow, metal microstructure and the nature of the initial colonizing species. When the biofilm covers an actively corroding point (or creates one), the SRB may be favored. This is due to the ability of SRB to utilize hydrogen generated at the cathode to metabolize a wide range of substrates. The SRB also have a high iron requirement; this element is obviously present at a corroding site.

As the SRB metabolize organic compounds and sulfate, sulfide is produced. This sulfide will depolarize the anode and precipitate iron sulfide (Figure 2). The consumption of hydrogen will depolarize the cathode and drive the corrosion reaction. The precipitated iron sulfide is cathodic to steel; this effect is a result of the facilitation of hydrogen reduction at the cathodic area [10]. It has been

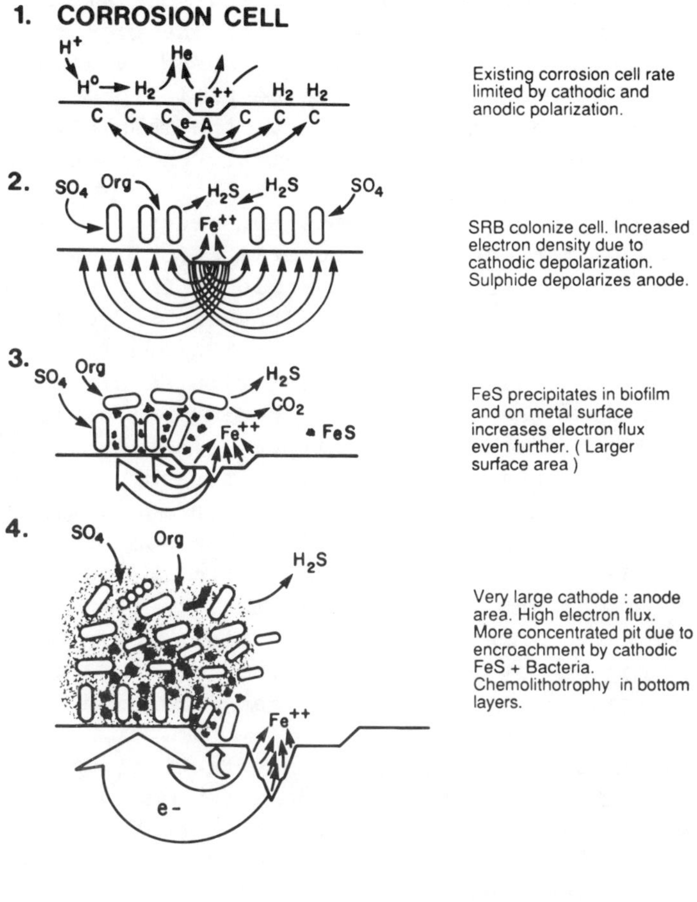

FIGURE 2. Sequence of corrosion events involving sulfate reducing bacteria.

postulated that the SRB interact with the iron sulfide to remove hydrogen from the deposit. The effective cathodic area is much increased and the depolarization effect is magnified. The mixture of bacteria, iron sulfide, and glycocalyx conductive to protons and corrosion current allows the development of a structured ecosystem where the bacteria nearest the flowing stream utilize nutrients from the environment. The metabolic by-products produced enter the glycocalyx and are utilized by SRB below the surface that have access to the hydrogen from the corrosion process. As we have seen, this hydrogen allows the metabolism of otherwise non-utilizable substrates such as acetate and CO_2 (Figure 1).

SRB corrosion can be rapid and severe. This is explained by the anode/cathode area effect. A typical SRB count in a corroding area would be in the order of 10^9 organisms/cm^2. These would have an effective surface area by themselves of over 100 cm^2 [14]. When these SRB are in intimate contact with finely divided iron sulfides, the effective cathodic area could be increased by an order of magnitude. This in itself will explain the severe corrosion induced by these bacteria: the cathodic area is 10^3 (or more) times the size of the anode. With a huge cathode and a small anode, the current flow is only limited by the ability of the SRB lower down in the biofilm to acquire sulfate. Even if those organisms lower down perish, the finely divided iron sulfide can still conduct protons to the metal surface and produce hydrogen.

The ability to use hydrogen effectively allows the SRB to flourish in a new ecosystem man has created for them: corroding metal surfaces. The factors that allow the SRB to be the prime agent of mineralization of organic matter and sulfate in nature allow them to dominate in the corrosion process.

Control Measures

Because they function within a biofilm on the metal surface, the detection and destruction of corrosion causing bacteria is a difficult task. Most industrial users rely on planktonic bacterial counts as an indicator of contamination and to monitor the effectiveness of control measures. In cases where the potential for microbial corrosion is great, it is obvious that monitoring microbial growth on surfaces is of paramount importance. In these cases, the use of the recently developed biofilm samplers is highly recommended [4].

Analytical methods for most of the corrosion causing bacteria are primitive when one considers the economic importance of their activities. In particular, the detection and enumeration of the iron fixing bacteria and the sulfate reducing bacteria is poorly developed. One study indicated that 99% of SRB present in samples were not detected!

Three general control techniques are available. These may be used in combination or separately.

Mechanical removal

Given that the driving force for many types of bacterial corrosion (especially the SRB) relies on the highly structured biofilm, scraping and pigging operations that disrupt the biofilm can reduce corrosion rates [4]. Experiences in seawater injection flowlines indicate that while many bacteria remain within the system, their corrosive effects are much reduced [17].

Nutrient removal

In many systems, flourishing populations of bacteria are maintained because of inadvertant supplies of an essential nutrient. The magnitude of a bacterial corrosion problem can often be reduced by identifying possible limiting nutrients. For example, some waters may be low in sulfate, but if an oxygen scavenger is added to control oxygen, 50-100 ppm of sulfate can be introduced into a system. Similarly, a surface water contaminated with bacteria but with low organic content (as is often the case) may be mixed with a sterile water rich in organic substrate (e.g. process water, oilfield water). The above scenarios can lead to a population explosion downstream of the mixing point.

Biocides

Chemical inhibition of bacteria is the method usually selected. Blind application of a material based on the supplier's recommendation may not ameliorate the problem. The following points should be considered for biocide treatments:

1) Make sure the monitoring program is effective.
2) Ensure that the biocide is soluble in the water used and active at the system pH and TDS.
3) Ensure that the biocide is compatible with other treatment chemicals.
4) Test the biocide against the organisms from the target system in the system water.
5) Test against sessile populations.
6) The higher the dosage, the better.
7) The longer the contact time, the better.
8) Application methods are critical--some systems may respond better to slug or continuous doses.

Case History of Corrosion Control Using Sessile Microorganism Sampling

Repeated corrosion failures in an oil collection pipeline system prompted a thorough study when traditional application of biocide protection failed to improve the situation. We reasoned that the system, which carries crude oil containing <1% water, was subject to bacterial corrosion by biofilm bacteria and that small "slug" doses of biocide were insufficient to kill the corrosion-causing, sulfate-reducing bacteria (SRB) within these protected adherent masses [15]. Accordingly, we resolved to install sessile sampling studs, identical to those used in the Robbins Device (see insert on inside back cover) [11], in rotatable spools (Figure 3) at strategic locations within the system. Our objective was to identify those locations at which potentially corrosive biofilm populations of SRBs had become established and to kill the sessile organisms with sufficiently high doses of effective biocides.

Oil was collected from 11 "ship out" tanks (1-11 in Fig. 3), all of which were owned by different combinations of operating companies and all of which collected oil from patterns of producer wells in West Central Alberta. The tanks were drained by pumps, via 8" vacuum lines, and the pump assemblies contained facilities for biocide injection. Relatively short branch 8" lines connected the ship out tanks to 4 main 8" lines (A-D in Figure 3) that carried the oil to a refinery complex. Spools containing sessile sampling studs were installed in all vacuum lines and in all 4 main lines (| | | in Figure 3) with rotatable bushings so that they would be located on the bottom of the pipe (6 o'clock position) during operation and rotated to the top of the pipe (12 o'clock position) for sampling. Planktonic "grab" samples could be taken during study withdrawal at any of these points, during momentary shut down of the line, and via drainage plugs in all of the ship out tanks.

After all of the spools, which were numbered S1-S11 in the individual vacuum lines and SA-SD in the 4 main lines, had been installed and "on line" for 2 months, a system-wide planktonic and sessile SRB sampling program was initiated. Planktonic samples for SRB were taken via the drainage plugs in all 11 ship out tanks and at the 4 main line spools (SA-SD), while sessile SRB samples were taken at each spool in the system. These data are entered on Figure 3, with notations for planktonic (P) and sessile (S) at the four locations where both types of samples were taken. At this time, an effective biocide was being continuously injected at low levels at each pump, which was downstream from the vacuum line spools in all cases.

The most startling and useful conclusion from these data is that the presence of planktonic SRB in a ship out tank does not, automatically, mean that potentially corrosive SRB populations will develop on the wall of the vacuum line pipes. This is illustrated in ship out tanks #1, 7 and 11. Conversely, a low planktonic count in the water component of the ship out tank oil does not guarantee that corrosive biofilms containing SRB will not develop on the vacuum line pipes. An example is found in ship out tank #8. The continuous injection of biocide at all ship out pumps obviously controlled both planktonic and sessile SRB growth in main lines A, C and D but plantonic bacteria had survived in line B and a small sessile population had begun to develop on studs in the spool (SB) in this line.

Because the local microbial ecology of some ship out lines had clearly not allowed the development of sessile SRB populations upstream of biocide injection, biocide treatment of these individual vacuum systems was discontinued. This decision constituted a considerable savings in biocide costs, and the presence of the main line spools gave us an opportunity to detect any loss of control of biofilm SRB that might have resulted from this decision.

In line D, biocide injection was discontinued at ship out tanks #7, 9 and 11, and maintained at ship out pumps #8 and 10. Continuous monitoring at the spool (SD) in the main line had not produced evidence

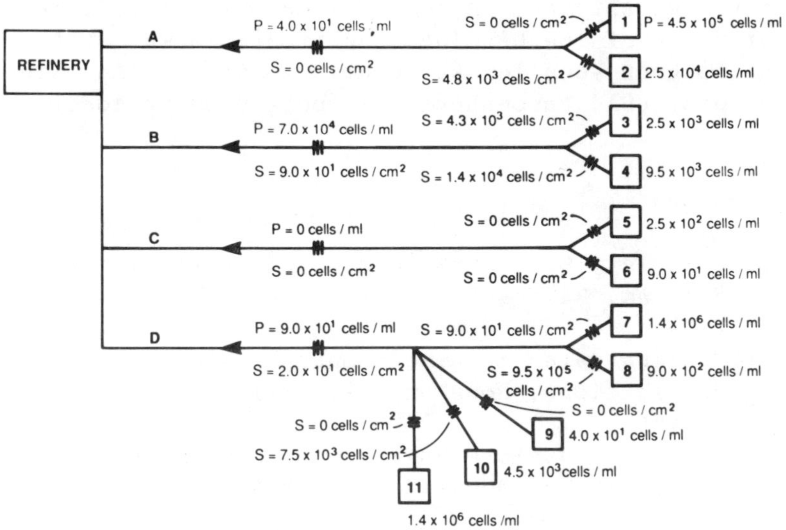

FIGURE 3. A schematic diagram of the oil collection system examined in this study. Main pipelines A-D drain ship out tanks 1-11 and the location of sessile sampling spools SA-SD and S1-S11 are shown by the symbol |||. Planktonic SRB counts at the initiation of the study are indicated by the prefix P and are given in CFU/ml while sessile SRB counts are indicated by the prefix S and are given in CFU/cm^2.

of sessile SRB development during 3 years of regular monitoring. Biocide treatment was discontinued at both ship out pumps (5 and 6) in line C because planktonic and sessile SRB counts were very low throughout the system. Even though biocide treatment was maintained in both ship out pumps (3 and 4) in line B, the sessile SRB count rose to 9.5×10^4 cells/cm^2 at spool SB three months after the monitoring program was instituted. This potentially dangerous problem was controlled by addition of high concentrations of biocide for an extended period of time. In line A, the maintenance of low biocide concentrations was sufficient to keep sessile SRB counts at zero at the main line spool (SA) throughout the monitoring period.

References

1. Brierley, C.L. 1982. Microbiological mining. Sci. Am. 247:44-53.

2. Cooling water treatment manual. NACE Publications, Houston, TX, 1971.

3. Corrosion basics--an introduction. NACE Publications, Houston, TX, 1984.

4. Costerton, J.W. and E.S. Lashen. 1984. Influence of biofilm on efficacy of biocides on corrosion-causing bacteria. Materials Performance 23(2):13-17.

5. Costerton, J.W., G.G. Geesey, and P.A. Jones. Bacterial biofilms in relation to internal corrosion monitoring and biocide strategies. Paper no. 54, Proceedings of the National Association of Corrosion Engineers annual meeting, NACE Publications, Houston, TX.

6. Dexter, S.C. (ed.). 1986. Biologically-induced corrosion. NACE Publications, Houston, TX.

7. Hamilton, W.A. 1983. The sulfate-reducing bacteria: their physiology and consequent ecology in microbial corrosion. The Metals

Society, London.

8. Lee, S.H., J.T. O'Conner, and S.K. Banerjee. 1980. Biologically mediated corrosion and its effect on water quality in distribution systems. J. Am. Water Works Assoc. 72(11):636-645.

9. Little, B. and P. Wagner. 1987. Microbiologically-induced cathodic depolarization. Paper no. 370, Proceedings of the National Association of Corrosion Engineers annual meeting, NACE Publications, Houston, TX.

10. Martin, R.L. and R.R. Annand. 1981. Accelerated corrosion of steel by suspended iron sulfides in brine. Paper no. 255, Proceedings of the National Association of Corrosion Engineers annual meeting, NACE Publications, Houston, TX

11. McCoy, W.F., J.D. Bryers, J. Robbins, and J.W. Costerton. 1981. Observations on biofilm formation. Can. J. Microbiol. 27:910-917.

12. Pankhania, I.P., Gow, L.A., and Hamilton, W.A. 1986. The effect of hydrogen on the growth of _D. vulgaris_ (Hildenborough) on lactate. J. Gen. Microbiol. 132:3349-3356.

13. Park, P.B. 1975. Biodeterioration in aircraft fuel systems. p. 105-126. _In_ R.J. Gilbert and D.W. Lovelock (ed.), Microbial aspects of the deterioration of materials, Academic Press, New York.

14. Postgate, J.R. 1984. The sulfate-reducing bacteria, 2nd ed. Cambridge University Press, Cambridge.

15. Ruseska, I., J. Robbins, J.W. Costerton and E.S.

Lashen. 1982. Biocide testing against corrosion-causing oilfield bacteria helps control plugging. Oil & Gas J. 80:253-264.

16. Schumaker, M. 1979. Effects of marine organisms. p. 366-387. In M. Schumaker, (ed.), Seawater corrosion handbook, Noyes Data Corp., New Jersey.

17. Surinach, P.P. 1987. A new concept of treating surfaces exposed to oilfield water systems. SPE 16262. San Antonio Conference.

18. Tatnall, R.E. 1981. Case histories: bacteria-induced corrosion. Paper no. 130, National Association of Corrosion Engineers annual meeting, NACE Publications, Houston, TX.

19. Wagner, P.A. and B.J. Little. 1986. Applications of a Technique for the Investigation of microbially-induced corrosion. Paper no. 121. National Association of Corrosion Engineers annual meeting, NACE Publications, Houston, TX.

CHAPTER 4. BACTERIAL ENDOTOXINS (PYROGENS) IN PURIFIED WATERS

Tom Novitsky, Associates of Cape Cod, Woods Hole, MA, 02543

Introduction

The water which covers 70% of our planet is an environment ideally suited to bacteria. Watson *et al*. [21] demonstrated that ocean water contained more than 10^4 bacteria per ml in the deepest ocean trenches and more than 10^6 bacteria per ml in coastal areas. Numerous investigators, using a combination of microscopic and chemical techniques, have confirmed these findings and extended them to freshwater ecosystems including wells, reservoirs, lakes, and streams. In purified waters, a similar situation exists; that is, bacterial communities occur and can even flourish. In many respects purified water is similar to an oligotrophic freshwater environment: the major difference being that bacteria that grow in purified water often result in biofouling.

One biochemical method used by Watson *et al*. [21] as an indirect determination of bacterial number and biomass in the aquatic environment was the *Limulus* amebocyte lysate (LAL) assay. This assay is a sensitive and quantitative assay for bacterial endotoxin. The LAL assay has provided investigators with the means for utilizing endotoxin measurements as an indicator of not only endotoxin but other bacterial related contaminants in purified water. Endotoxins are associated with Gram-negative bacteria, which comprise over 90% of the total microbial population in purified waters [15]. In this chapter, the role of endotoxins as contaminants in pure water systems will be discussed.

Nature of Endotoxins

Endotoxins were first described by Pfeiffer in 1892 [17]. As the name implies, endotoxins are part of the bacterial cell (envelope) and exhibit toxic properties when injected into warm-blooded animals. The toxicity takes the form of fever, shock, and even death depending on the concentration and type of endotoxin injected. Because endotoxins are the most common cause of fever in humans, they are often referred to as pyrogens, especially in the pharmaceutical industry.

The chemical nature of endotoxin was first elucidated by Boivin *et al*. in 1932 [1]. The portion of the endotoxin molecule possessing the most potent and diverse biological activity was later described by Westphal *et al*. in 1952 [22]. The term lipopolysaccharide (LPS) is the term used for this active portion of endotoxin. Although LPS has been studied intensely since its discovery, only recently has the structure of the biologically active portion of the molecule been completely characterized [19].

LPS is an amphiphilic molecule composed of three distinct regions: the O-polysaccharide (somatic antigen), the core, and the lipid A. The O-polysaccharide imparts hydrophilicity to the molecule and is composed of repeating di- and tri-saccharide subunits. The O-polysaccharide is species specific, with small changes in the number and placement of subunits responsible for various serotypes within a bacterial species. The core region acts as a bridge between the hydrophilic O-polysaccharide and the hydrophobic lipid A. The core region consists of heptose sugars and ketodeoxyoctulonic acid (KDO), and is fairly similar within bacterial genera. The hydrophobic lipid A is the biologically active portion of the molecule. It is responsible for inducing fever and for reactivity with LAL. Lipid A is comprised of glycosidically-linked glucosamines substituted with phosphates and fatty acids. Lipid A chemistry is believed to be highly conserved within families of bacteria, although only the enteric bacteria have thus far been studied in detail.

Endotoxins or lipopolysaccharides are found in nature as an integral part of viable and non-viable bacteria (bound LPS) as well as free in the surrounding media (free LPS). It was once believed that a Gram-negative bacterium needed to lyse or die before LPS was liberated into the surrounding medium (free LPS). This is not the case. Bacteria release various amounts of LPS during growth; the amount released being dependent on the species, growth conditions, etc. It has also been shown that relatively mild chemical and mechanical treatment, often not sufficient to kill the bacteria, can cause the release of LPS [7].

Sources and Implications of Endotoxins in Water Systems

Endotoxins are of particular concern to pharmaceutical manufacturers that utilize purified water or water for injection (WFI) to prepare injectable drugs or devices used to deliver these drugs. Since endotoxin can cause physical harm to humans and animals, the United States Food and Drug Administration closely regulates the level of allowable endotoxin in drug products, biologicals, and medical devices. The USP has recently set a limit of 0.25 EU/ml for sterile water for injection and sterile water for irrigation and a limit of 0.5 EU/ml for bacteriostatic water for injection and water for inhalation [20]. An endotoxin unit, EU, is roughly equivalent in potency to 0.1 ng/ml of an *Escherichia coli* endotoxin standard.

As was stated in the introduction, bacteria are ubiquitous in aquatic ecosystems. In surface water, the influx of nutrients from runoff as well as by-products of photosynthesis adequately support large heterotrophic bacterial communities. Some photosynthetic bacteria and cyanobacteria (both contain endotoxin) can also be quite abundant. Well water on the other hand, unless contaminated from a surface source (e.g., from a septic system), usually contains few bacteria. Most water supply systems treat water to eliminate pathogenic bacteria. These treatments do not necessarily kill all the bacteria present, nor do they destroy endotoxins. Thus, feed water for a water purification system usually has some bacterial loading and almost always contains substantial amounts of endotoxin. Typical city tap water contains from 1 to 100 EU/ml. Much of this endotoxin is free, having been released during previous bacterial growth or after bacteria were killed during water treatment. The remainder of the endotoxin is associated with intact viable and non-viable bacteria. There is a growing body of opinion that most intact bacteria observed in aquatic ecosystems where there is no evidence of disinfection are indeed viable even though they may be impossible to culture. On the other hand, a lysed bacterium is obviously non-viable. This question of viable vs. non-viable becomes of prime concern in the interpretation of direct count and LPS data.

Depending on the water supply, bacterial loading as well as endotoxin levels are quite variable. In most cases, however, endotoxin levels are usually high--regardless of the number of intact or culturable bacteria present in the supply.

Detection of Endotoxins

Prior to the acceptance of the LAL test, rabbits were routinely used to determine the pyrogenicity (ability to cause fever) of drugs and devices. Today, some pharmaceutical manufacturers who have not yet validated their personnel in the use of LAL still employ the rabbit pyrogen test. In fact, the rabbit test, or USP pyrogen test [20], has served the industry well for over forty years, making pyrogenic episodes resulting from contaminated intravenous drugs almost a rarity in the United States. The USP pyrogen test, however, lacks sensitivity (see Table 1), is at best semiquantitative, is time and sample consuming, and is expensive. All of these factors, as well as animal welfare groups, have hastened the acceptance of the LAL assay as an alternative pyrogen test procedure. Although LAL reacts only with bacterial endotoxin pyrogen (non-endotoxin pyrogens exist and can be quite common under certain circumstances), a vast amount of data collected by the US Food and Drug Administration and the pharmaceutical industry since

TABLE 1. Detection limits of various endotoxin assays

Test	Sensitivity (EU/ml)
USP Pyrogen Test	0.5[1]
Gel-Clot LAL Assay	0.03
Chromogenic LAL Assay	0.01
Turbidimetric (kinetic) LAL Assay	0.001
Other chemical methods	>10

[1]Indicates the dose where 50% of rabbits fail test. Injection is usually 10 ml/kg. Dose of endotoxin would be about 5.0 EU/kg.

the introduction of the LAL test indicate the most prevalent pyrogen in pharmaceutical preparations (and especially purified water) is endotoxin. It is difficult for someone in the semiconductor industry, in basic research, or involved in water purification in general to understand the usefulness of an assay for endotoxin other than in the context of the aforementioned tests for pyrogens.

Several methods including the LAL assay have been used to measure LPS. Since LPS contains some unique constituents, including KDO and β-hydroxymyristic acid, traditional wet chemical assays can be used. Detection levels, however, are usually in the microgram range and natural samples often contain large quantities of non-LPS material which necessitate modifications of these assays. HPLC and TLC have been used on the intact LPS molecule but these methods lack sensitivity and accuracy. Antibodies to LPS (O-polysaccharide) have been used successfully in clinical applications to detect bacteria and free LPS. The extreme specificity of antibodies however, precludes their use in systems where a wide variety of bacterial species would be expected. Recently, antibodies to the lipid A and core regions of LPS have demonstrated greater cross-reactivity and thus may be capable of LPS detection from larger groups. These antibodies are still very limited to clinical applications. It is unlikely that the core and lipid A regions of LPS from bacterial species found in water will be similar enough chemically (antigenically) to allow use of an antibody assay for endotoxin in purified water.

Unlike the chemical and antibody assays, the LAL test is highly specific for LPS and is extremely sensitive. Although the LAL is activated by the lipid A portion of LPS, variations in the chemical nature of the lipid A between bacterial groups do not seem to significantly affect the reaction. Quantitative differences between endotoxin species have been documented [5], but the differences are small and may be insignificant especially when the LAL test is used on samples containing endotoxin from a bacterial community or mixed culture. Watson _et al_. [21] demonstrated conclusively that the LAL assay could be used to determine bacterial number in an aquatic environment. Figure 1 illustrates the relationship between cell number and bacteria-associated (bound) LPS. Watson *et al*. correlated bound LPS concentration to be related to bacterial abundance as well [15].

LAL is an extract of the blood cells or amebocytes of the North American horseshoe crab, *Limulus polyphemus*. That the extract reacts with minute amounts of endotoxin forming a turbid solution or gel-clot was elucidated in two papers by Drs. Levin and Bang in 1964 [9,10]. The LAL reagent consists of a pro-clotting enzyme, a clotting protein (coagulogen), and inorganic salts (required for optimal activity). LAL is usually provided freeze-dried and requires reconstitution with endotoxin-free water. In the most commonly used LAL assay, the gel-clot test, reconstituted LAL reagent is added to an equal volume of liquid sample in a 10 x 75 mm endotoxin-free glass test tube. The mixture is incubated undisturbed at $37 \pm 0.5^{o}C$ for 60 ± 2 minutes. Following incubation, the test is scored by inverting the reaction tube 180 degrees. A positive result is obtained if a solid gel-clot is formed; that is, one that withstands the inversion. A negative result is scored if the gel-clot breaks upon inversion.

For most applications in the pharmaceutical industry, the gel-clot test is performed. Pharmaceutical interest in LAL is based on the requirement for the absence of endotoxins (pyrogens). Thus in the pharmaceutical industry, a negative result (indicating non-pyrogenicity) is desired. In the pharmaceutical industry the sensitivity of the LAL is selected to correspond to a predetermined level, usually a few fold more sensitive than that required to detect a pyrogenic level of endotoxin. A somewhat different approach is used to measure the endotoxin content of a solution. In this case, a quantitative method such as one described by Watson *et al*. [21] or Remillard *et* al. [18] may be used. These LAL methods require the construction of a standard curve which spans the range of endotoxin concentrations to be measured. Samples containing unknown amounts of endotoxin are tested and results are obtained by extrapolation from the standard.

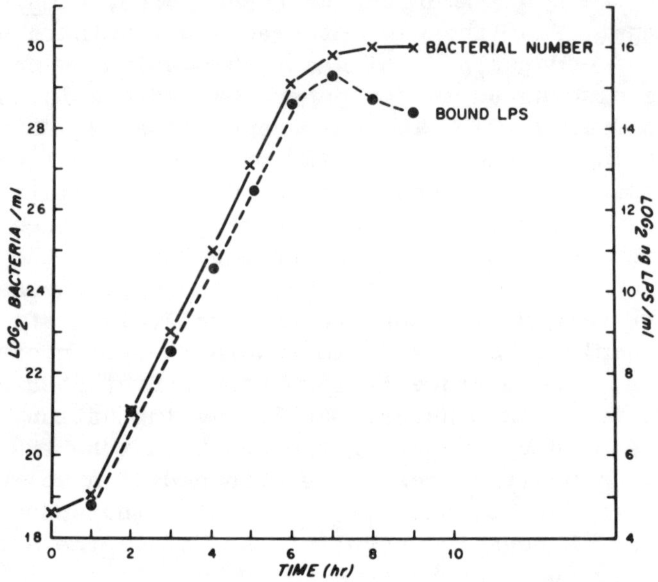

FIGURE 1. Correlation of LPS as determined with the LAL test and bacterial number as determined by fluorescence direct count epifluorescence microscopy.

Basically, there are two quantitative LAL methods, the turbidimetric assay (kinetic or end-point), and the chromogenic assay (end-point). In the turbidimetric assay, an increase in turbidity (leading up to a gel-clot) is measured as either the rate of increase (kinetic) or the final optical density (end-point) after a fixed period of incubation. Either the rate function or the optical density can be plotted against endotoxin concentration to generate a standard curve. In the chromogenic assay, a synthetic substrate replaces the native coagulogen in the LAL reagent. This substrate consists of a small peptide containing the amino acid linkage of the coagulogen that is acted upon by the activated clotting enzyme and a para-nitroanilide chromophore. When the endotoxin-activated clotting enzyme cleaves the peptide bond in the substrate, the chromophore becomes yellow and can be measured with a spectrophotometer. As in the end-point turbidimetric assay, a standard curve can be constructed by plotting optical density vs. endotoxin concentration. For studies attempting to correlate other bacterial parameters with endotoxin, it is imperative to use one of the quantitative LAL assays.

As with other enzyme assays, the LAL test is subject to interferences. These interfering factors can be physical (e.g. temperature, vibration, viscosity) or chemical (e.g. pH, osmolarity, presence of chelating agents) in nature (Table 2.). Interference with the LAL test is usually one of degree. The most common way to eliminate interference is by dilution of the sample to diminish the effects of the interfering substance while retaining sufficient endotoxin for detection.

Endotoxins as Indicators of Biofouling

Although the LAL test can be used as an indirect test for the determination of bacterial number (Figure 2), bacteria in pure water systems may be predominantly attached to surfaces. As it has been explained elsewhere in this book, most bacteria are attached to surfaces. Endotoxins also adsorb to surfaces as they are quite "sticky". Although it is possible to measure endotoxin adsorbed to surfaces [3], it can be quite resistant to removal as was demonstrated by

Novitsky *et al*. [16].

Measurement of endotoxin in a water system therefore usually involves only circulating endotoxin [14]. It is not known whether there is a good correlation between circulating endotoxin and endotoxin which is a part of the biofilm. Of course, if we assume that all endotoxin measured in the system (minus that from the feed) originates *de novo* in the system, then circulating endotoxin may correlate with the degree of biofouling.

TABLE 2. General classes of LAL interfering factors

Type	Optimum
Physical	
Temperature	32-40°C (USP requires 37°C)
Viscosity	1.4 centipoises @ 35°C
Vibration	minimal for formation of gel-clot
Chemical	
pH	6 - 8
Alcohol	< 0.5% w/v
Chelating agents	< 0.01 M divalent ion equivalents
Urea	< 1 M
Osmolarity	< 300 milliosmoles/liter
Other	as determined by experiment

Under certain conditions, the amount of circulating endotoxins will remain relatively constant or increase at a slow but steady rate. These conditions are indicative of a system operating at an optimal level. That is to say, carbon beds and deionizers are functioning and not over colonized, and likewise with filters and other components of the system. As carbon beds and deionizers are fouled and as bacteria increase on the surface of the piping and other system components, the amount of endotoxin "released" into the system increases. This increase can be sudden and dramatic as threshold levels of bacteria are reached in certain components. Dramatic increases in circulating endotoxin have also been noted following perturbation of the system. For example, disinfection of carbon beds (steam treatment) can result in the release of large amounts of endotoxin into the circulating water. The actual amount is related to the

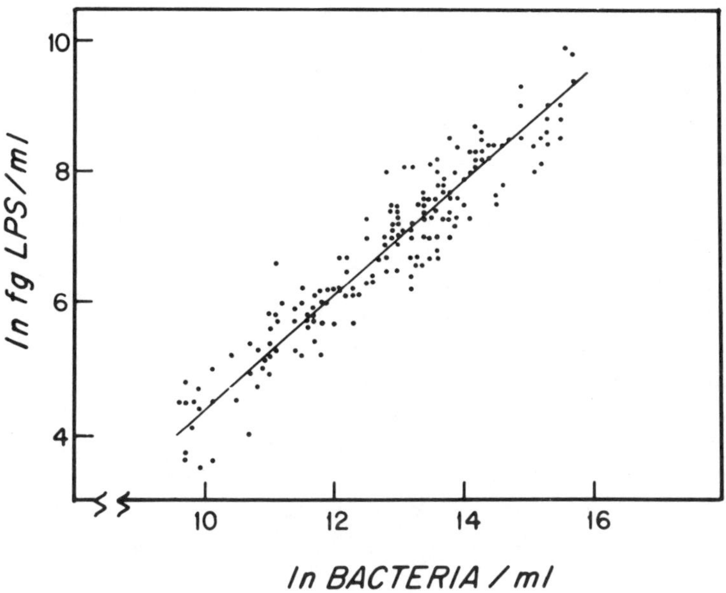

FIGURE 2. Correlation of bound LPS as determined by the LAL assay and bacterial number as determined by direct count epifluorescence microscopy using a laboratory culture.

degree of fouling prior to disinfection. These kinds of increases are usually not well correlated with increased bacteria levels. Although the monitoring of endotoxin to evaluate disinfection programs is not common, a report by Chitty [2], describes a sucessful application of the LAL test in prescribing a biocide dosing schedule for a cooling tower. Use of the LAL test in this regard resulted in a reduction in the frequency of biocide addition and monetary savings.

Endotoxin Correlation with Other System Chemistry

Reports in the literature indicate that endotoxin concentration correlates well with bacterial numbers determined by cultural or direct microscopic techniques [4,15,21]. An example of the correlation between plate count and free and bound LPS for a laboratory bacterial culture is shown in Figure 3. Determinations based on direct counts, LPS, and total organic carbon (TOC) are also well correlated [15]. These correlations are dependent, however, upon the lability of the organics present in the system. It is not surprising that bacterial number correlates with TOC since organic carbon can serve as a colonization stratum for bacteria and bacteria are actively involved in recycling organic carbon. In one report, TOC was shown to correlate with viable counts [13]. A data summary from Novitsky *et al*. [15] is shown in Table 3.

Control of Endotoxins in Pure Water Systems

Water purification systems serve two purposes in industrial water systems; they lower levels of chemical impurities (organic and inorganic) and they eliminate (or lower) levels of viable bacteria. Unfortunately, methods and devices used to achieve these goals may, under certain conditions, actually add to the contamination of the system. Since pure water systems and their functional components are described in detail elsewhere in this book, this discussion will be limited to generation and removal of endotoxin by system components themselves and by procedures used to disinfect or maintain these components.

Distillation has long been used for the production of pure water. The pharmaceutical industry has successfully employed distillation to prepare chemically- (endotoxin free) and biologically-pure

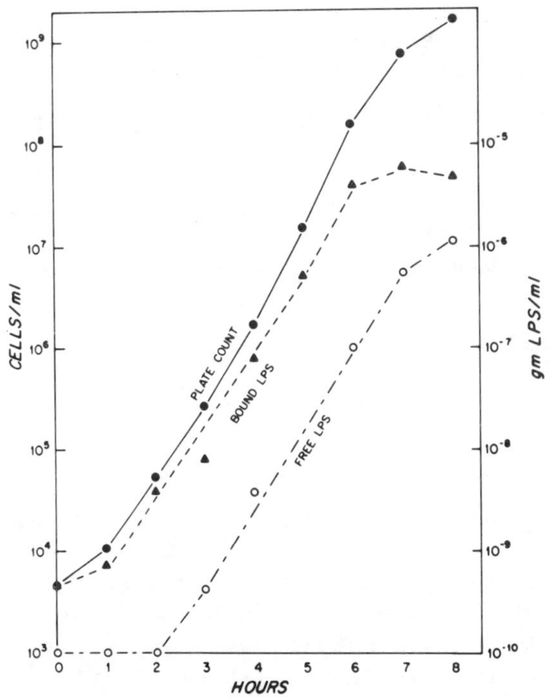

FIGURE 3. Correlation of colony forming units with bound and free LPS.

water [20]. Distillation systems containing a baffle located between the boiler and the condenser are extremely efficient at eliminating endotoxins, bacteria, and other particulates. Endotoxin or bacterial problems with stills are rare and usually occur only if water droplets (rather than water vapor) reach the condenser [11]. This problem can be further exacerbated if the boiler water contains extremely high endotoxin levels.

Filtration methods for the removal of endotoxins can be very efficient [Table 4]. Reverse osmosis and ultrafiltration are also very effective at removing

TABLE 3. Correlations between direct counts of bacteria, total organic carbon, and endotoxin

Statistical Parameter	Direct Count vs. Endotoxin	Direct Count vs. TOC	Endotoxin vs. TOC
Correlation Coefficient[1]	0.886	0.683	0.629
Sample Number	44	37	34

[1] P>0.0001

bacteria and endotoxin from water. It is generally agreed that an ultrafilter rated at 20,000 daltons or less will effectively remove most endotoxins. It should be noted, however, that under conditions of high loading, some endotoxin species of lower molecular weight may pass through these filters. Lipid A, for example, can easily pass a 5000 dalton cutoff ultrafilter and is readily reactive with LAL.

Carbon beds also remove endotoxins, although the mechanism is one of adsorption. Endotoxins can also be removed by adsorption to anionic and cationic exchangers, the surfaces of reverse osmosis and ultrafilter membranes, the walls of pipes and storage containers, and also by the surfaces of membrane filters. Filters possessing a zeta potential efficiently remove endotoxins although ordinary depth filters and 0.2 micron rated membrane filters can remove substantial amounts under certain conditions.

Softeners do not remove endotoxin. However, because of their chemical nature (containing high salt concentrations), the growth of bacteria is not encouraged. One problem related to softeners is that softened water tends to foam in distillation systems operated under suboptimal conditions. The foam and entrained particulates (e.g. bacteria and endotoxins) can be carried over to the condenser.

Unfortunately, the majority of components in a water system, although designed for chemical removal, can and do with time become fouled and can add contaminants--usually organic carbon--to the system. When colonization of membrane filters, carbon beds, and deionizers reaches a certain level, these components become bacterial and/or endotoxin "factories" adding contaminants to the system and possibly promoting biofouling of other parts of the system.

The addition of ultraviolet lamps to pure water systems can control bacteria but probably does little to remove endotoxin. For the effective destruction of endotoxin in water systems, chemical methods as described in the next section are required.

Chemical Removal of Endotoxins

Several chemical disinfection methods for pure water systems have the potential for destroying endotoxins. The most common disinfectant, chlorine, destroys little endotoxin. As chlorine kills bacteria, endotoxin is actually added to the system. Hydrogen

TABLE 4. Relative removal efficiencies (in descending order) for endotoxins in water

Method	Conditions for Complete Removal[1]
Distillation (including thermo-compression)	Used according to manufacturer's specifications. System must include baffling system.
Reverse Osmosis	Used according to manufacturer's specifications. Systems tend to foul and must be monitored and sanitized regularly.
Ultrafiltration	Plate or hollow fiber units at 5,000 to 20,000 MW rating, depending on source and nature of endotoxin.
Charge Modified Media Filtration[2]	Used according to manufacturer's specifications.
Microporous Membrane Filtration[2]	Used according to manufacturer's specifications; <0.2 μm, hydrophobic.
Microporous Membrane Filtration	Used according to manufacturer's specifications, <0.2 μm, hydrophilic.

[1]Complete = final water product contains less than 0.001 EU/ml.

[2]Depending on conditions (i.e., pH, resistivity, etc.), relative removal efficiencies for these two membrane filtration methods may be reversed.

peroxide (H_2O_2), on the other hand, can efficiently destroy endotoxin [6]. Although elevated temperatures greatly increases the efficiency of H_2O_2, increasing the concentration above 3% provides little additional benefit [6]. Recently, claims have been made that ozone (O_3) is effective in reducing endotoxin [8]. Although, ozone, theoretically, should be quite capable of destroying endotoxin, little emperical data is available to evaluate its activity. Acids and bases can also destroy and/or help eliminate endotoxin by hydrolysis, disaggregation, and surface removal [12]. Treatment with acid/base during the regeneration of deionizers undoubtedly lowers the endotoxin content. Other chemical disinfecting agents employing a combination of ingredients have also been shown to be effective in lowering or removing endotoxin. These agents usually contain chlorine and one or more surface active agents. There is presently no data available on the destruction or removal of endotoxin by quaternary ammonium compounds.

Summary

Endotoxin, a component of Gram-negative bacteria, is a ubiquitous macromolecule present in natural and pure water systems. Detection and quantitation of endotoxin is relatively easy using the *Limulus* amebocyte lysate (LAL) assay. The presence and amount of endotoxin can be used to assess the quality and/or operational efficiency of a water system. Increased levels of endotoxins with time during the operation of a pure water system are indicative of bacterial colonization and can therefore be a useful predictor of bacterial contamination and biofouling.

References

1. Boivin, A. and L. Mesrobeanu. 1933. Contribution a l'etude de la composition chimique des bacteries. Les substances phosphorees au cours de l'autolyse bacterienne. C.R. Soc. Biol. 112:611.

2. Chitty, N. 1985. A key to preventing Legionnaire's disease. Laboratory Sci. Technol. 1:15-16.

3. Dexter, S.C., J.D. Sullivan, Jr., J. Williams III, and S.W. Watson. 1975. Influence of substrate wettability on the attachment of marine bacteria to various surfaces. Appl. Microbiol. 30:298-308.

4. Evans, T.M., J.E. Schillinger, and T.G. Stuart. 1978. Rapid determination of bacteriological water quality by using Limulus lysate. Appl. Environ. Microbiol. 35:376-382.

5. Firca, J.R. and J.A. Rudbach. 1982. Reference endotoxin: A practical rationale. p. 121-130. In S.W. Watson, J. Levin, and T.J. Novitsky (ed.), Endotoxins and their detection with the Limulus amebocyte lysate test. Alan R. Liss, Inc., New York.

6. Gould, M.J. and T.J. Novitsky. 1985. Depyrogenation by hydrogen peroxide. p. 84-92. In Technical Report No. 7, Depyrogenation. Parenteral Drug Association, Inc., Philadelphia.

7. Jorgenson, J.H., and R.F. Smith. 1974. Measurement of bound and free endotoxin by the Limulus assay. Proc. Soc. Biol. Med. 146:1024-1031.

8. Khoudary, N. 1985. The use of ozone in the production of ultrapure water. Pharm. Engin. 5:30.

9. Levin, J. and F. Bang. 1964. The role of endotoxin in the extracellular coagulation of Limulus blood. Bull. Johns Hopkins Hosp. 115:265-270.

10. Levin, J. and F. Bang. 1964. A description of cellular coagulation in the Limulus. Bull. Johns Hopkins Hosp. 115:337-356.

11. Mahoney, R.F. 1985. Depyrogenation by distillation. p. 37-44. In Technical Report No. 7, Depyrogenation. Parenteral Drug Association, Inc., Philadelphia.

12. McCullough, K.Z. and T.J. Novitsky. 1985. Detoxification of endotoxin by acid and base. p. 78-83. In Technical Report No. 7, Depyrogenation. Parenteral Drug Association, Inc., Philadelphia.

13. Mittelman, M.W. 1985. Biological fouling of purified water systems: Part I, Bacterial growth and replication. Microcontamination 3:51-70.

14. Novitsky, T.J. 1984. Monitoring and validation of high purity water systems with the Limulus amebocyte lysate test for pyrogens. Pharm. Engin. 4:21-33.

15. Novitsky, T.J., M.E. Dawson, and W.V. Collentro. 1985. A new approach for monitoring bacteria: The kinetic LAL assay. Ultrapure Water November-December p. 10-12.

16. Novitsky, T.J., J. Schmidt-Gengenbach, and J.F.

Remillard. 1986. Factors affecting recovery of endotoxin adsorbed to container surfaces. J. Parent. Sci. Technol. 40:284-286.

17. Pfeiffer, R. 1892. Untersuchungen uber das Choleragift 11:393-412.

18. Remillard, J.F., M.C. Gould, P.F. Roslansky, and T.J. Novitsky. 1987. Quantitation of endotoxin in products using the LAL kinetic turbidimetric assay. p. 197-210. In S.W. Watson, J. Levin, and T.J. Novitsky (ed.), Detection of bacterial endotoxins with the Limulus amebocyte lysate test. Alan R. Liss, Inc., New York.

19. Rietschel, E.Th., H. Brade, L. Brade, K. Brandenburg, U. Schade, U. Seydel, U. Zahringer, C. Galanos, O. Luderitz, O. Westphal, H. Labischinski, S. Kusumoto, and T. Shiba. 1987. Lipid A, the endotoxic center of bacterial lipopolysaccharides: relation of chemical structure to biological activity, p. 25-55. In S.W. Watson, J. Levin, and T.J. Novitsky, (ed), Detection of bacterial endotoxins with the Limulus amebocyte lysate test. Alan R. Liss, Inc., New York.

20. United States Pharmacopeial Convention, Inc. 1985. U.S. Pharmacopeia, XXI ed., p. 1181-1182. Mack Printing Co., Easton, Pennsylvania.

21. Watson, S.W., T.J. Novitsky, H.L. Quinby, and F.W. Valois. 1977. Determination of bacterial number and biomass in the marine environment. Appl. Environ. Microbiol. 33:940-946.

22. Westphal, O., O. Luderitz, and F. Bister. 1952. Uber die extraction von bacterien mit phenol/wasser. Z. Naturforsch. Teil B7:148-155.

CHAPTER 5. ULTRAVIOLET AND OZONE SYSTEMS

Theodore H. Meltzer, Capitola Consulting Company, Arlington, VA 22202

Rip G. Rice, Rice International Consulting Enterprises, Ashton, MD 20861

Introduction

Fouling miroorganisms can be considered as viable, organic, self-replicating contaminants in industrial waters systems. Chemical oxidation reactions can inactivate microorganisms. However, the non-viable residues of microrganisms such as endotoxins, can also interfere with the water product systems and various industrial and consumer products. Some of these may be removed by filtration, distillation, and other treatments such as chemical oxidation. One purpose of ultraviolet (UV) radiation, and ozone treatment is to effect such disinfection and/or oxidative degradations.

The actions of ozone and/or UV radiation often are through the agency of free radicals, produced by or from the incorporation of these two treatment processes. Two covalently joined atoms can separate and produce moieties which then possess one bonding electron each. The resulting entities, neutral in electrical charge, are called free radicals. The impelling drive of free radicals to acquire another electron usually makes them chemically aggressive. A gamut of free radical activity exists, the more reactive ones enter into reactions with their neighboring organic (or sometimes inorganic) molecules. Hence their durations are much shorter than they would be in pure solution, free of reactive contaminants. Ozone itself can react with most organic molecules. However the hydroxyl free radical, which can be one product of ultraviolet emanations and of ozone in aqueous solutions, especially in the presence of hydrogen peroxide, is most efficacious in degrading organic molecules by chemical oxidation.

Table 1 presents a comparison of different oxidants and their oxidizing potentials. It is seen that ozone is a very powerful oxidizing agent; in fact, the most powerful readily available for water treatment. By the same token, the hydroxyl free radical is seen to be even more powerful an oxidizing agent.

TABLE 1. Relative Oxidation Power of Various Oxidizing Species

Species	Oxidation Potential Volts	Relative Oxidation Power[1]
Fluorine	3.06	2.25
Hydroxyl free radical	2.80	2.05
Atomic oxygen	2.42	1.78
Ozone	2.07	1.52
Hydrogen peroxide	1.77	1.30
Perhydroxyl free radicals	1.70	1.25
Hypochlorous acid	1.49	1.10
Chlorine	1.36	1.00

[1]based on chlorine as reference (= 1.00)

Ultraviolet and Ozone Systems

Ozone

Properties of Ozone

Ordinary molecular oxygen consists of two atoms of oxygen in peaceful chemical union. Normally quite stable, molecular oxygen can be decomposed through the agency of UV radiation or the application of high electrical energy into two reactive oxygen atoms or free radicals. In aqueous solutions, the interaction of atomic oxygen with water molecules may give rise to two hydroxyl free radicals:

$$O\cdot \; + \; H\text{-}O\text{-}H \rightarrow (HO)\cdot \; + \; \cdot(OH)$$

Ozone is an allotrope of oxygen consisting of three oxygen atoms. It is relatively unstable. Liquid ozone (100% pure) is readily exploded. Gaseous compositions containing ozone/oxygen mixtures in excess of 20% constitute potentially explosive mixtures. However, in commercial generation of ozone from air or from oxygen, concentrations above 10% ozone are difficult to prepare. At such concentrations, gaseous compositions containing ozone have not been reported to be explosive. The decomposition product of gaseous ozone is oxygen. Certain catalysts, as well as organic matter with which ozone can react, facilitate its decomposition to elemental oxygen.

Gaseous ozone is only sparingly soluble in water, in general about 13 times more soluble than oxygen. Its specific solubility is dictated by Henry's Law, which relates solubility of a partially soluble gas in a liquid to its partial pressure above that liquid. Different liquid media have different Henry's Law constants.

Chemical reaction rates with ozone usually bear some relationship to reagent concentration, as well as temperature, pH, and the chemical nature of the oxidizable solutes and inorganic impurities. Therefore, efforts sometimes are made to combine the action of ozone with UV radiation or/and hydrogen peroxide so that the more reactive hydroxyl free radicals may be produced in high concentrations.

The half-life of ozone in once-distilled water is about 25 minutes at 20°C. However, in the presence of impurities with which it can react, it is considerably less. The ready destruction of ozone can be beneficial. UV radiation at a wavelength of 254 nm converts it into oxygen and hydroxyl free radicals. This can be accomplished conveniently in a recirculating loop; the water being released upon signal from an ozone detector. The rate of ozone decomposition in water is pH-dependent. At pH 8, typical of that of many drinking water supplies, ozone has a half-life of much less than 30 minutes. Its decomposition rate, initiated by hydroxyl ions, increases with pH [11].

Comparisons With Chlorine

The control of microorganisms in industrial water systems has been managed by the use of numerous disinfectants or sterilants. Each has been shown to have its merits and limitations. Inevitably, ozone comes to be compared with chlorine for this type of application. Chlorine is a relatively stable compound, and hence is not described by any half-life characteristic. The relatively short half-life of ozone in water means that significant concentrations of dissolved, residual ozone probably will not endure throughout an extensive water distribution system. The microbial population, controlled to that point, will begin to flourish again. Long-lived chlorine would not be so fugitive. On the other hand, once chlorine has served its useful purpose it (and many of its by-products) is more difficult to remove. Adsorption by activated carbon or reaction with bisulfite or other reducing solutions are the usual means, and each has its complications [20].

Chlorine may convert many organic substances into trihalomethanes, and other halogenated derivatives. Some of these have been associated with carcinogenicity. The short half-life of ozone means that upon discharge, treated waters are less likely to be toxic to aquatic life. Indeed, the decomposition serves to increase the dissolved oxygen level of the water being treated. This is usually desirable; however, increased corrosion may also result.

It has been cited that ozone kills the bacterium, Escherichia coli 3,125 faster than the same molar concentration of chlorine [27]. Table 2 compares the bactericidal efficiency of ozone with that of other chlorine disinfectants [23]. The ozone residual levels represented range from 0.0005 to 0.5 mg/l [27]. In this case, the action of ozone may be so rapid that

TABLE 2. Specific lethality coefficients at 5°C[1]

Agent	Enteric Bacteria	Amoebic Cysts	Viruses	Spores
ozone (O_3)	500	0.5	5	2
HOCl as Cl_2	20	0.05	1.0 up	0.05
OCl^- as Cl_2	0.2	0.0005	<0.02	<0.0005
NH_2Cl as Cl_2	0.1	0.02	0.005	0.001

[1] values refer to specific susceptibility or lethality coefficient, K, expressed as $[mg/l]^{-1} \times min^{-1}$ at 5°C.

contact times may pose no problems, assuming that efficient transfer of ozone to the aqueous phase has been effected. When ozone is added to water using an efficiently operated in-line static mixer, bacteria are killed within seconds [27,28]. Chlorine requires diffusion through the cell walls of the organisms in order to degrade enzymes. Ozone lyses the cell walls directly. In disinfection applications, ozone is more powerful than chlorine by factors of from 5 to 1,000, depending upon the form in which the chlorine is dispensed as well as the type of organisms present. Chlorine as hypochlorous acid, the gas dissolved in water at pH at or below 7, is the most effective form of chlorine; chloramine is the least, and hypochlorite ion (present at pH above 7.0) is of intermediate germicidal activity.

Action Upon Microorganisms

The presence of viruses in waters is of considerable current concern. Enteroviruses and viruses of infectious hepatitis can exist for long periods, even in well-maintained potable water distribution systems. Ozone is far more effective than UV radiation or chlorine in inactivating such viruses. For example, complete inactivation of polio type viruses was effected within 30 seconds using a dissolved ozone residual of less than 0.5 mg/l [17]. French public health officials [5,6] specify ozone concentrations of 0.4 mg/l maintained over a minimum contact period of 4 minutes, to guarantee a 99.9% inactivation of Poliovirus Types I, II, and III (Table 3).

The concentrations of chlorine required to inactivate *Giardia lamblia* cysts are higher than those recommended in water treatments for coliform bacteria. Comparisons based upon 99% inactivation levels of cysts of *Giardia lamblia* and *Giardia muris* indicate that they are more sensitive to ozone than are cysts of *Entrobacter histolytica* or *Naegleria gruberi*. *Giardia* cysts have been reported to be more resistant to ozone than are *E. coli*. In France, where ozone treatment is used widely for primary disinfection of potable water supplies, the residual ozone level is maintained at 0.4 mg/l for 8-12 minutes. Such a level is sufficient to inactivate in excess of 99% *Giardia* cysts, even at low temperatures. Wickramanayake *et al*. [37] showed that the contact times required to inactivate two logs of *G. lamblia* cysts with 0.15 mg/l residual ozone at pH 7 are 1 and 4 minutes at 25° and 5°C, respectively. Cysts of *G. muris* are 1.5 and 3.5 times more resistant to ozone at 25° and 5°C, respectively.

TABLE 3. Lethality coefficients (K) for ozone at pH 7 and 10-15°C

Organism	K	C_t99:10[a]
Escherichia coli	4600	0.0010
Streptococcus faecalis	3000	0.0015
Mycobacterium tuberculosis	1000	0.0500
Poliovirus	460	0.0100
Bacillus megatherium (spores)	150	0.1000
Entamoeba histolytica	50	0.0300

[a] C_t99:10 = concentration in mg/l for 99% destruction in 10 min.

Ozone Generation

Nearly all large scale, commercial ozone generators employ the corona discharge principle. Dried air, or oxygen itself, is passed between a high voltage and ground electrode separated by a dielectric material (Figure 1). Considerable electrical energy is required for formation of the ozone-producing electrical discharge field. A minimal energy expenditure requires the use of dry air (maximum dew point of minus 60°C). One to two percent ozone is produced in dry air. For the same electrical energy applied to the generator, double that concentration of ozone is obtained using dry oxygen In excess of 80% of the applied energy is converted to heat which, if not rapidly removed, causes the ozone produced to decompose, particularly above 35°C (95°F). Proper cooling of the ozone generator is crucial to maintaining consistent ozone yields.

Ozone generator design alternatives, as well as considerations of using oxygen rather than air are discussed by Rice & Bollyky [30]. Means of dispersing the ozone, whether by diffuser, injector, static mixer, etc., are discussed by these authors, as is the treatment of ozone-containing contactor off-gases. In older ozonation plants, the contactor off-gases were diluted with air and discharged through tall stacks. Modern practices, however, include the injection of the ozone-containing off-gases into raw waters having high ozone demand, thermal destruction at 260-316°C, passage through towers of moist granular activated carbon (GAC), reduction with a solution of alkaline thiosulfate, passage over a catalytic manganese dioxide-containing formulation, or exposure to UV radiation.

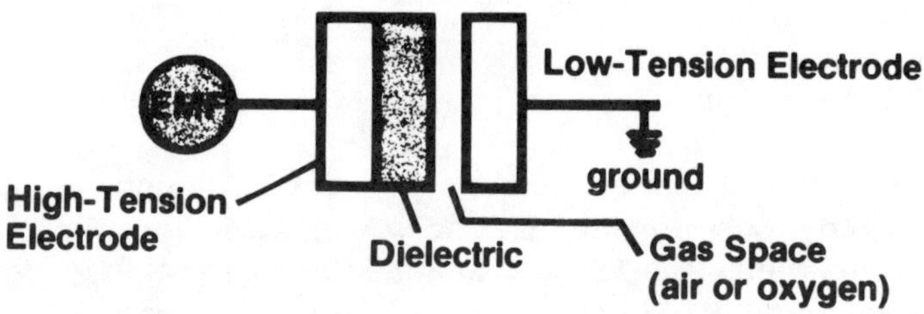

FIGURE 1. Diagram of ozone generator based on corona discharge principle.

Ozone Generation by UV Radiation

Ozone also can be generated in oxygen or air by UV radiation itself, although only in relatively low concentrations per unit time; insufficient to treat large volumes of water economically. Thus, more than 44 kW are required to generate 1 kg of ozone from dried air under high gas flow rates [26] by UV, compared with 12-17 kW by corona discharge.

The oxidative efects of UV radiation on aqueous solutions are often ascribed to the presence of ozone. The more likely explanation is generation of hydroxyl free radicals in solution via the UV radiation. Ozone is probably not involved, except to provide some of the feed source of the hydroxyl free radicals.

Ozone Dosage

Pseudomonas spp., *Staphylococcus*, spp. *Candida* spp., *Penicillum* spp, *Proteus* spp., *Clostridium* spp., *Schistosoma* spp., *Salmonella* spp., *Shigella* spp., *Vibrio cholerae*, and *Entamoeba* cysts are readily destroyed by ozone in aqueous suspensions. Ozone sterilization is a very rapid first order reaction. Vegetative bacteria at concentrations of 10^6-10^7/ml are totally destroyed at ozone concentrations of 0.01-1.0 mg/l within five minutes. *E. coli*, *Streptococcus* spp., and *Bacillus* spp. are inactivated at ozone concentrations of less than 0.2 mg/l within 30 seconds. Bacterial spores are 10 to 15 times more resistant. *Bacillus* and *Clostridium* spores at levels of 10^7/ml require higher ozone doses and contact times for sterilization. Ozone concentrations of 0.2-0.5 mg/l and contact times of 10 minutes are recommended for water sterilization where large bacterial spore populations are involved. Viruses are extremely susceptible to ozone. Residual ozone concentrations of 0.1 to 0.8 mg/l and contact times of less than 30 seconds suffice for Polio, Encephalomyocarditis, Coxsackie, Enterovirus, Vasicular stomatitis, GDVII, Rhabdovirus, and Herpes viruses [13].

Analysis For Ozone

Oxidation of iodide ion by ozone to release elemental iodine which, in turn, produces a deep blue color in the presence of starch, and that is titratable by standardized sodium thiosulfate solution is a very convenient method for measuring dissolved ozone concentrations. However, this method is not specific for ozone alone. Other materials capable of oxidizing iodide ion to iodine interfere; e.g., organic peroxides, hypochlorite ion, chlorine dioxide, hydrogen peroxide, etc. It has been widely employed for several decades for ozone analysis, however, for its convenience. Nevertheless, its use no longer is recommended [12].

Ultraviolet absorption is selective for ozone at 253.7 nm, in the liquid or gaseous phases. Single or double beam spectrophotometers may be used; the more expensive double beam instrument more conveniently compares the water to be treated with that containing ozone. Otherwise, blank determinations must be made to corect for background effects. The primary limitation of UV absorption spectroscopy is the current lack of agreement as to the value of the molar extinction coefficient to be employed. Thus, the accuracy of this method can vary by about 30% in either direction, depending upon which value of the molar extinction coefficient is taken [12].

An amperometric method based on the electrochemical conversion of ozone to oxygen is also in use in many ozonation facilities to measure levels of dissolved ozone. The current required to effect the conversion is proportional to the dissolved ozone concentration. The sensitivity of this method is good. Its chief limitations are the potential fouling of the electrodes by organic contaminants, scale-forming entities, sulfides, and the fact that calibrations are based on the iodometric method.

Decolorization of indigo disulfonate solutions by

ozone is the basis for the determination of ozone which is subject to fewer interferences than any of the previously described procedures. Further discussion of analytical methods for determining ozone in the gaseous or water phases can be found in a recent work by Rice *et al*. [32].

Operational Considerations

The ozone generated for biofouling control can be effective only to the extent that it dissolves in the conveying water. In accordance with Henry's Law, ozone concentration varies directly with its partial pressure, and inversely with temperature. Therefore, it is recommended that the generated ozone be injected into the water stream after the circulation pump and before the heat exchanger. The dynamic head of the heat exchanger and the subsequent piping, as well as the static head of the riser pipe to the tower spray nozzles (Figure 2), act to increase the ozone solubility at the injection point. The lower temperatures upstream of the heat exchanger favor higher ozone solubilities. The ozone residual levels will be reduced at the heat exchanger and at the tower fill by air stripping, by increases in the water temperature, and by the decrease in its pressure at the spray nozzles. The quantity of ozone injected to achieve a desired residual level should reflect these influences.

The type of ozone contactor used with cooling towers may differ in individual locations. The location of the ozone generator, space restraints, and rate of water flow will influence the choice. If the cooling water is recycled at sufficiently high speeds, static mixers, spray driers, or injectors can be utilized with no additional input electrical energy.

Cooling towers are engineered so that the cascading water droplets exchange their unwanted heat with an upward cooling draft of air. Roughly 72 cubic feet of air are swept through every gallon of water every minute. The water contains the ozone, the air does not. Therefore, the air current serves to strip the water of its ozone content. Additionally, the forced air draft introduces new populations of microorganisms into the aqueous system. A substantial decline in ozone content occurs in the descending

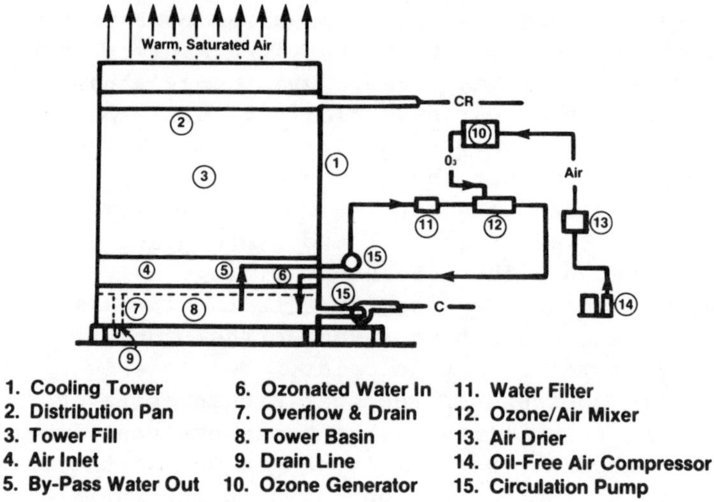

FIGURE 2: Diagram of water cooling system indicating location of ozone generator.

waters. This becomes progressive as the remaining ozone encounters and reacts with the microorganisms harbored within the cooling tower fill. As a result, the cooling tower sump contains water with the lowest concentration of ozone in the system. Assessment of the sufficiency of ozone residual concentrations should be made at this location. The sump is dynamic, recirculating at one to two minute intervals.

Generally five to ten minutes of contact time are desired for total control of microorganisms in circulating water when residual ozone levels of 0.4 mg/l are involved. If even as low an ozone residual as 0.02 mg/l continuously bathes the cooling tower fill and sump, microbial control should be satisfactory, provided a prolonged, nearly infinite contact time is allowed in the cooling tower. However, to attain such an ozone residual concentration, an 0.4 mg/l residual level must be present in the circulating waters. At these residual ozone concentrations, the microbes suspended in the circulating waters are eliminated within one to two minutes. Those adhering to the tower-fill surfaces or resident within the tower sump are held to a bacteriostatic condition by the prolonged exposures to lower ozone concentrations present.

Ultraviolet Radiation

It has been known for over a century that the UV wavelengths present in sunlight have germicidal effects. The practical utility of UV radiation for the control of microorganisms followed the devising of the mercury vapor lamp. This form of radiant energy induces photochemical reactions involving biologically active molecules in microorganisms. The resulting molecular alterations inhibit growth, and in higher doses, result in cellular death.

The UV light spectrum is broad enough to include wavelengths of unequal effects upon organisms. Even for efficacious wavelengths, the influence of the UV radiation may differ. Also, the devices designed to discharge UV light are not all alike in the radiant energy they generate. Moreover, the UV emanations are susceptible to absorption by the molecules of the organism-suspending medium. Their germicidal influence may become attenuated by the time they reach the microbe. That UV radiation is of practical utility in controlling most organisms is beyond challenge. Where organisms are killed by UV radiation, the subsequent removal of dead microbes may be required, as by filtration. In addition to their role as particulate contaminants, UV-inactivated microorganisms can be a source of organic carbon (TOC) and pyrogenic lipopolysaccharides.

Generation of UV Radiation

The most common method of generating UV radiation is by discharge lamps. Typically, a glass or quartz tube contains an inert gas plus a metal, usually mercury. Electrodes are located at each end of the tube. They may be cold or preheated. In either case, an ignition voltage is required to begin the discharge. Its minimum value is dependent upon the length of the tube, (i.e., the distance between the electrodes), the internal pressure of the tube, and its contents. Ignition voltage may range from 20 V to several thousand volts. Preheated electrodes, promotive of ions, electrons, and electron carriers, require lower ignition voltages. If the lamp burning voltage is less than half that of the line voltage, then the lamp can be started easily by means of ballast and starter. Otherwise a stray field transformer is needed. Lamps based on mercury/argon mixtures exhibit the highest efficiency. Mercury lamps are designated as high or low pressure. The partial pressure within the discharge tube will depend upon the lowest temperature along the tube, a cool spot. Typically, the pressure may range from 0.025 pa at 0^oC, to 0.8 pa at 40^oC, to 10^5 pa (atmospheric pressure) at 360^oC, to pa values above 10^7 above 800^oC. The mercury vapor pressure will be from 0.001 to 0.1 atm Hg in the low, and from 0.1 to 10 atm Hg in the high regions.

UV Wavelengths

During the discharge, particularly at low pressures, the mercury atoms in their excited states radiate mainly at wavelengths of 184.9 nm, the resonance line, and 253.7 nm, the intercombination line. This results in a large generation of the short UV lines. Depending largely upon the current density and the tube cross-section, the efficiency of this type generation may be above 60%. Whether the 185 nm line can penetrate the bulb depends upon its materials of construction. The use of quartz tubes permits the passage of lower wavelengths.

Increasing the pressure within the tube, i.e., the density, promotes collisions of mercury molecules. These tend to decrease the excited states incidence with concomitant increases in energetically lower lying states; i.e., the longer wavelengths become stronger. At typical medium pressures of 10^5 pa, about 2% of the input power is radiated at 254 nm, and about 10% at wavelengths below 300 nm. At higher pressures, the radiations below 260 nm become absorbed, and the longer wavelengths broaden into a continuum.

UV radiation can, for practical purposes, be divided into four sections: UV-A between 400 and 315 nm, UV-B between 315 and 280 nm, UV-C covering from about 280 to 200 nm, and vacuum-UV, below 200 nm. The last is strongly absorbed by air. UV-C is used primarily to destroy microorganisms. The 253.7 nm line emitted by low pressure mercury discharge lamps can inactivate microorganisms such as protozoa, bacteria, molds, yeasts, viruses, fungi, and algae. It is this type of lamp that is used primarily in the UV disinfection of water. The broad band emanations of medium-pressured mercury lamps are also suitable for UV germicidal applications.

Germicidal UV Spectra

The germicidal effectiveness of UV radiation is dependent upon its wavelength. Different organisms show a slightly different sensitivity to various parts of the UV spectrum. However, an average sensitivity can be characterized. Figure 3 presents the germicidal action curve, roughly Gaussian, wherein relative germicidal effectiveness is plotted against wavelength. The action maximum is at about 265 nm. The 253.7 nm line is about 0.85 times as effective. It corresponds to a conversion of 35% of the input power. Table 4 illustrates the relative germicidal effectiveness of various UV wavelengths [22].

The disinfection action is dependent not only upon the UV emission spectrum, but also upon the radiation intensity, the duration of exposure, the sensitivity of the microorganisms involved, and the UV-transmission of the suspending medium. Clear air offers little impediment to transmission. However, dissolved substances even in clear water can reduce the radiation intensity by adsorption. Iron and manganese compounds above 1 mg/l and 0.2 mg/l, respectively, sugars, and humic acids have such effects [10]. In defining the extent of transmission by a given solution, the measurement is made at 754 nm wavelength, and the depth of water, whether 10 or 100 mm, is stipulated.

Ultrasonic effects promote a more efficient UV germicidal action, possibly by promoting the separation of individual organisms from one another and/or from sheltering particulate matter, thus exposing them more fully.

UV Dosage

UV devices are rated by their capacities to treat water at specific flow rates. Actually, the UV dosage is the real consideration. It is the product of radiation intensity and the exposure time, and is expressed as microwatts-seconds per square centimeter ($\mu Ws/cm^2$). Restated, it is a quantity of energy, the irradiance in the case of UV radiation, in watts/m^2 multiplied by time in seconds, yielding joules/m^2. It is the dosage that defines the effectivity of the UV experience, not the watt input or radiated output of the UV lamps.

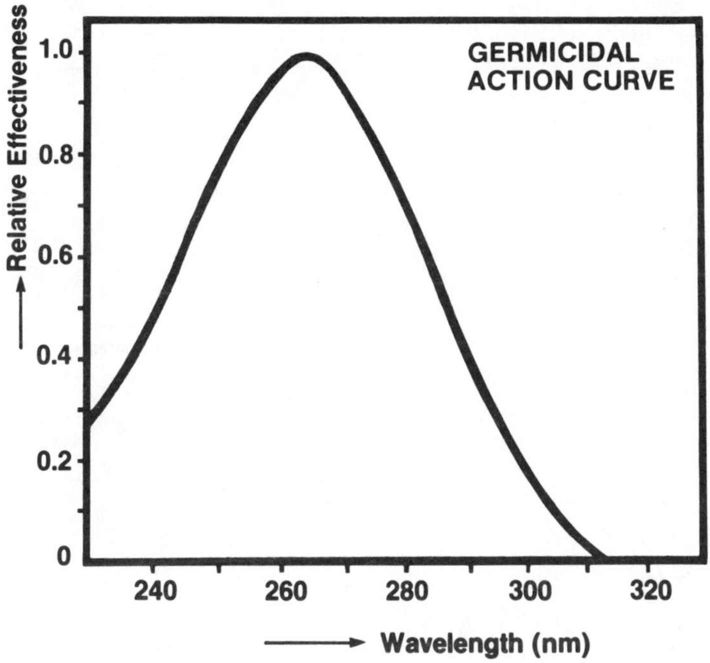

FIGURE 3. Germicidal action curve in the ultraviolet region of the energy spectrum.

Ellner and Ellner [8] have devised a biological method for measuring the actual dosages delivered by ultraviolet devices. A standard curve is constructed by plotting test bacterial survivors against graded UV radiation doses. Water seeded with the test organism

TABLE 4. Effectiveness of UV radiation as a function of the wavelength

wavelength in nm	relative germicidal effectiveness
210	0.02
215	0.06
220	0.12
225	0.18
230	0.26
235	0.36
240	0.47
245	0.61
250	0.75
255	0.88
260	0.97
265	1.00
270	0.93
275	0.83
280	0.72
285	0.58
290	0.45
295	0.31
300	0.18
305	0.10
310	0.05
315	0

is treated in the device. By reference to the standard curve, the effluent cultured for survivors reveals the actual delivered dose.

There is no possibility of overdosage by UV irradiation, in terms of negative effects, except as they relate to economic considerations of overdesign. In the treatment of potable waters, a minimum dose of 25 mJ/cm^2 is both effective and economical. In the beverage industry, the minimum dosage has been established at 50 mJ/cm^2; in ultrapure water treatment, 100 mJ/cm^2.

Antopol and Ellner [3] reported that Legionella pneumophilia, the causative agent of Legionnaires Disease which is found in cooling towers of air conditioners, is inactivated by UV light. Approximately 380 $\mu W/cm^2$ killed 50%; 920 uW/cm^2, 90%; 1840 $\mu W/cm^2$, 99% and 2760 $\mu W/cm^2$ 99.9% of the test population of microorganisms. UV dosages required to produce 90% inhibition of E. coli, Salmonella typhi, Serratia marcescens, and Pseudomonas aeruginosa are 2110, 2140, 2200, and 5500 $\mu W/cm^2$ respectively [32].

The radiation output of a UV lamp depends upon the lifespan of the device. This is not constant, and may also vary from lamp to lamp. Guarantees of stipulated service levels differ from 500 to 8,000 hours. Beyond this duration, the performance of the lamp is undependable, and the lamp should be replaced.

The radiation exposure time is a function of the water flow velocity and the geometry of the radiation chamber. The organisms suspended in the conveying fluid must reside in the radiation zone long enough to absorb a minimum lethal dose. If the suspending medium has too rapid a velocity through the chamber, the minimum dosage may not be achieved. The flow pattern is of importance as well. Most UV radiation generators are quartz tubes that are centrally mounted as a core within the radiation chamber, usually a sturdy, pressurized steel tube. The flowing water surrounds the light tube. This minimizes the thickness of the flow channels, promoting penetration by the UV radiation. Random flow is promoted by the turbulence occasioned by baffles. This minimizes the distance between the UV tube and any organisms in the radiation chamber.

The quartz tube that protects the UV lamp should be cleaned at least each time the UV bulb is replaced. In applications in which slimes are built up, such as in fish hatchery waters, the tubes are fitted with automatic wipers. These serve to maintain the constancy of UV transmission.

Mode of Action of UV Radiation

Chemical disinfectants initially attack the outer surfaces of organisms. Hence, considerable contact time may be required. However, UV radiation penetrates the microbial cell wall or cell membrane, being absorbed by DNA, RNA, and enzyme components. Cell morphology has a direct bearing on the dosage required for inactivation. For example, Gram-positive bacteria which have thicker outer cell walls, are more difficult to destroy than are the Gram-negative bacteria. Spores with their thicker walls are more resistant than vegetative cells. Algae are 100 times more resistant than are bacteria. Various microorganisms differ in their resistance to UV radiation. Environment, the stage of growth, and nutrition may affect resistance. Thus, in water, about 30 J/m^2 of UV radiation is required to kill *E. coli* bacteria to an extent of about 90%. Lower doses are needed to kill organisms in air [10]. Giardiasis is an intestinal disease that is encountered more and more frequently in the United States. It arises from waterborne agents, namely cysts of the pathogenic protozoan *Giardia lamblia*. Rice & Hoff [31] found that UV radiation was far less effective against *G. lamblia* than against *E. coli*. Even a dosage of 63,000 $\mu W\text{-}s/cm^2$, four times the minimum U.S. Public Health Service recommendation, was inadequate to inactivate *G. lamblia* cysts in water.

Microbes exposed to UV radiation are not all killed or inactivated at once; a constant fraction of the living number dies in each increment of time [15]. The survival ratio, that fraction of the initial number surviving at any given time, can be approximated by an exponential function of the product of the radiation intensity and the exposure time. Plots of this ratio are straight lines on semilog paper. This means that to decrease the survival ratio from 0.1 to 0.01, i.e., one order of magnitude, double the dose of UV radiation is required. If a survival ratio of 0.00001 is needed, the dose corresponding to the survival ratio of 0.1 must be increased five-fold. Table 5 lists the 0.1 survival dosages for a number of organisms.

Influencing Factors

Certain commercial designations require clarification. UV lamps with double intensities do not necessarily mean double dosages, unless the flow capacity is identical. One UV generator designed for treating 100 gpm volumes and having twice the intensity of a unit intended for 50 gpm of water flow does not yield double the dosage of that smaller generator. All else being equal, the dosages could be the same. Interested readers may wish to refer to Abshire's observations concerning various manufacturer's claims [1].

It is not enough to speak of percentage reductions in microbial concentrations. If the raw water contained 10^6 cfu in a specific volume, then a 99.99% reduction would mean that 10^2 cfu still are present in that water volume--a very high number. In designing a UV treatment unit, the number of permissible surviving organisms should be stipulated as well. The treatment system then can be designed not to yield populations in excess of this number.

The U.S. Environmental Protection Agency, Office of Drinking Water is advocating new disinfection regulations for assuring that water supplies obtained from surface waters are free of *Giardia lamblia* cysts and enteric viruses. In these new regulations (to be promulgated), although the *goal* for *Giardia* cysts and viruses is zero, EPA will be requiring a minimum 3-log

TABLE 5. Approximate dose values for 1 survival ratio of 0.1 of various microorganisms at 253.7 nm [18]

Bacterium	Dosage	Bacterium	Dosage
Bacillus anthracis	45	Pseudomonas aeruginosa	55
B. megatherium (veg)	11	P. fluorescens	35
B. megatherium (spores)	27	Salmonella enteritidis	40
B. parathyphosus	32	S. typhosa - Typhoid fever	22
B. subtilus	70	S. paratyphi-Enteric fever	32
B. subtilus (spores)	120	S. typhimurium 80	
Micrococcus luteus	197	Proteus vulgaris	26
Serratia marcescens	24	Vibrio comma - Cholera	34
Clostridium tetani	130	Shigela dysenteriae	
Corynebacterium diphtheriae	34	Dysentery form	22
Eberthelia typhosa	21	Shigella flexneri	
Escherichia coli	30	Dysentery form	17
Leptospira sp.:		Shigella paradysenteriae	17
infectious jaundice	32	Spirillum rubrum	44
Micrococcus candidus	61	Staphylococcus albus	18
Micrococcus piltonencis	81	Staphylococcus aureus	26
Micrococcus sphaeroides	100	Streptococcus hemolyticus	22
Mycobacterium tuberculosis	62	Streptococcus lactis	62
Neisseria catarrhalis	44	Streptococcus viridans	20
Phytomonas tumefaciens	44	Mycobacterium tuberculosis	100

reduction in Giardia levels (99.9%) and a minimum 4-log reduction in pathogenic virus concentrations (99.99%) by whatever technique is utilized.

Selected Applications

Ozone, UV radiation, and their combinations with catalytically synergistic agents, such as hydrogen peroxide, are applied to the control of biofouling problems in many settings, not the least of these being potable water supplies. Three industrial applications are described below: cooling towers for air conditioning, usages in the pharmaceuticals industry, and removal of trace organic materials from ultrapure waters intended for integrated circuit electronic parts rinsing.

Seawater Cooling Systems

Heat exchangers and cooling towers provide environments wherein solid surfaces become subject to biofouling. The volumes of cooling water required may vary enormously, and may have very different origins and dissimilar microbial populations. Biofouling of heat exchangers, often manifested by slime formation on the inside of condenser tubes, results in a reduction of heat transfer efficiency, in corrosion of metals, and in a diminution of water flow rates. In large electric power plants, biofouling results in significant, recurring economic losses to the power industry.

Various chemical additives have been employed to minimize and/or prevent biofouling; chlorine being chief among them. Ozone has also been used to achieve this goal, and studies often have shown it to be technologically superior. Sugam and Guerra [35] conducted a comparative study at an electrical utility power generating station on a polluted brackish water. Chlorine was found to be more effective on a weight basis, but ozone could be used at a level less injurious to the indigenous aquatic species. This advantage could be heightened by intermittent use of ozone as the biocide. Advantages accrue to ozone from its scale-inhibiting action in addition to its biocidal effects [21].

However, the complexities of seawater offer special problems. Thus, the presence of bromide ion (up to approximately 65 mg/l) consumes ozone (as well as chlorine), and is oxidized to hypobromite and bromate ions. The hypobromite ions quickly establish an equilibrium with hypobromous acid ("free bromine"). Since the normal dosages of ozone or chlorine are considerably less than 5 mg/l, the bromide ion is present in considerable excess (up to 65 mg/l). Consequently, both ozone and chlorine generate "free bromine" in the seawater, which serves as the biofouling control agent. One is led to the obvious conclusion that it is more economical to generate free bromine from elemental chlorine than from ozone.

Majumdar & Glassman [18] reported that ozone is a more powerful oxidant/biocide than chlorine, that it leaves no undesirable residuals, that it imparts neither taste nor color to treated waters, that its required dosages need not harm aquatic life, and, importantly, that it does not increase the inorganic salt content of treated waters. However, most of the virtues claimed for ozone, while true for fresh waters, may not apply as well to seawater and other bromide ion containing waters. For while residual ozone may decompose into harmless (and even beneficial) oxygen, the products of its oxidative reactions (hypobromite and bromate ions) may be persistent and even toxic to some forms of aquatic organisms [4]. This same consideration applies to chlorine, which also oxidizes bromide ions to the same products.

Air Conditioning Cooling Towers

Studies have been made of the current use of ozone in cooling water towers of air conditioning systems at four locations in California [21]. In physical appearance, the ozone-treated towers generally were good, scaling and biological growth being limited to the outer edges in areas alternately wetting and drying. The waters generally were clean, but sometimes were colored. The ozone dose was estimated at a maximum of about 0.03 mg/l, based on circulating water flow. No heat transference problems were evident. Biofouling was judged to be controlled by the ozone. The heat exchange surfaces were not observed directly, but the waters appeared clear; their plate counts were relatively low at 7,000 to 54,000 colonies/ml. The cooling towers were comparatively clean and the air conditioners operated well.

Large quantities of ozone may have to be introduced, particularly in older cooling towers, where scale disintegration and the oxidative removal of accumulations of organic material and dried slimes will be necessitated. The initial high demand for ozone serves to preclean the cooling tower. Some quantity of ozone must be expected to be lost by decomposition, occasioned by elevated water temperatures, particularly above 80 ^{o}F (26°C). A dissolved ozone concentration of at least 0.5 mg/l immediately after the water exits the ozone contactor will suffice for cooling waters of older towers [34].

An installation in Alhambra, CA consists of a high frequency ozone generator, including an air predrying unit, of 90 g/h ozone generation capacity. Thus far, an 8-month operation has been flawless. It provides ozone for a model 372-102 Marley cooling tower servicing condenser water for 1,236 tons of air conditioning. The water is clear and odor-free. Its usage has declined to an average of 900 gal/day from 2,300 gal/day. The condenser pressure has undergone a corresponding reduction of two pounds. Inspection of the condenser showed it to be sludge-free with only a slight scale formation. The ozonation system is operated on a 20-minute off/on cycle. Plans are to prolong the off-stage to 30 minutes. The tower is inspected daily. Originally, the operation was planned for eight hours; it actually operates for only three hours. A 65% cost savings is projected. This is attributed by Stopka [34] to the 2% (by weight) concentration of ozone in air generated, coupled with a highly efficient ozone contactor.

An installation in Los Angeles based on the same

type of 90 g/h ozone generator utilizes two independent ozone contactors, each supplying a 2-inch, in-line static mixer contactor. The basins of each of two 750 ton cooling towers are supplied separately with water that is ozonized. Algae and scale disintegrated after eight days of ozone treatment. The water, previously dirty and yellow in color, is now crystal clear. There has been no need to add water treatment chemicals, and hence no need to remove their accumulation by any blowdown. The planned treatment cycle was decreased by 50%. A single 90 g/h ozone generator is planned to service independently four different cooling towers grouped together; each of 2,000 gallons of water using four sets of pumps, and 2-inch, in-line static mixer contactors. It is planned to alternate ozonation of successive towers automatically on a 30-minute cycle. Capital savings are projected from the use of one, rather than four separate ozone generators.

Three of four cooling towers (two 400 tons; one 300 tons) in a German dairy have been operating successfully with ozonation for over one year. The water here is freshwater, to which about 15 mg/l of bromide ion has been added. About 25% of the total volume of water recirculating in the three towers (25 m^3/h) is ozonized continuously at an applied ozone dosage of approximately 2 mg/l, then mixed with the balance of the cooling water (75 m^3/h: total applied ozone dosage = 0.5 mg/l).

During ozonation, many of the organics are oxidized. In addition, the added bromide ion is also oxidized to produce hypobromous acid, which flows throughout the cooling water system in the three cooling towers. Because there is always an excess of bromide ion over dissolved ozone, there is no ozone in the cooling towers themselves. However, the high concentration of ozone generated in the dried air (up to 3.5% by weight) oxidizes many of the contaminating organic components. The more stable HOBr maintains the heat transfer surfaces free of scaling and corrosion activities. In this manner, in which only 25% of the total cooling water is ozonized, less ozone is required to produce a high quality of cooling water and to prevent scaling and corrosion, without the use of added chemicals, other than bromide ion.

A Japanese team [24] has developed an unique system (MABOS) for generating ozone and applying it intermittently to cooling waters. In this system, ozone is generated from pure oxygen, then adsorbed onto silica gel. This allows the oxygen to be rerouted back to the ozone generator without recleaning or redrying. When ozone is required, it is simply desorbed from the silica gel by heating and then drawing the gas into the water by suction. Several major advantages accrue to this new system. First, generating ozone from pure oxygen allows smaller generators to be used than would be required using air; thus, capital costs are lower. Second, the oxygen gas exiting the adsorption columns can be recycled to produce ozone. Only minimal amounts of oxygen are therefore consumed. Finally, when ozone is desorbed and added to the water, only ozone is added to the cooling water--no oxygen is added above that which is formed as the solubilized ozone decomposes.

The original description of the MABOS system [24] demonstrated its application to fresh water cooling towers. A later description [26] demonstrated its use for seawater cooling water circuits. The system applies high concentrations of ozone to the cooling water once each day for 5-10 minutes. This prevents algae and slimes from attaching to the condenser tube walls.

A recent German modification of the MABOS concept [16] is called SORBOZON. Following the two silica gel ozone adsorbers is an "ozone surge tank", into which ozone desorbed from the two adsorbers is collected. From this surge tank, ozone can be fed continuously to the contacting system.

UV Sterilization in the Pharmaceuticals Industry

Abshire [1] investigated three different types of UV lamps in a study aimed at evaluating UV radiation as a sterilant for the pharmaceutical industry. He found the method applicable. Bacillus pumilis in its spore form, being relatively resistant, serves as a microbiological indicator for the validation of the UV irradiative process. Abshire [2] also utilized spores of Bacillus pumilis as a biological indicator in the UV sterilization of translucent polyethylene bottles. Ninety percent of the spores at a population level of 10^6 were consistently destroyed in two minutes using minimal doses of 360,000 ergs/mm^2 per 30 minutes. No spores survived a 10 min exposure. A minimal dose of 252,000 ergs/mm^2 per 30 min gave a D-value of four minutes. After 21 minutes, there were no surviving spores from a 10^6 population in the 2 ml aqueous test suspension. Table 6 lists the organisms tested by Abshire and details their UV resistances in terms of the UV doses in ergs/mm^2. The measured doses compare with those listed elsewhere in the literature. Table 7 from Abshire presents a classification of microbial resistance to UV irradiation in terms of the dosage required. Table 8 presents the author's findings relative to the decimal reduction times of selected organism strains following exposure to a UV intensity of only 100 μW/cm^2. The use of 100 μW/cm^2 intensities rather than 1,000 μW/cm^2 was made in order to allow more accurate calculation of D-values. Survival times were only on the order of seconds at the higher doses. Micrococcus radiodurans ATCC 13939 was the most resistant of the vegetative forms studied. Candida albicans ATCC 10231 also was resistant. The other organisms studied, including those of specific interest in the manufacture of ophthalmics, were sensitive to UV irradiation; most of the D-values were between one and two minutes.

TABLE 6. UV resistance data; Alcon's dose results compared to literature references

Organism[1]	UV Doses (ergs/mm²) Alcon	Literature	Author(s)
Micrococcus radiodurans ATCC 13939	14,400-57,600[2]	16,000 (99.999%)[3]	Bowling, Setlow
Micrococus luteus ATCC 9341	2,400-3,600	3,000 (99.0%)	Rubbo, Gardner
Escherichia coli ATCC 23224	120-800	462 (99.999%)	Greenberg, Karrer
Bacillus cereus ATCC 11778	1,200-1,800	No Reference Found	
Bacillus subtilis NCTC 2479	Not Tested	40,000 (99.99%)	Rubbo, Gardner
Bacillus subtilis (niger) ATCC 9372	57,600-72,000	No Reference Found	
Bacillus pumilus[4] BMT-18	36,000-72,000	No Reference Found	
Bacillus stearothermophilus ATCC 7950	14,400-28,000	No Reference Found	
Saccharomyces cerevisieae DJ-52875	2,000-3,600	68,000 (99.99%)	Rubbo, Gardner

[1]Particular strains listed are those studied at Alcon. All Bacillus strains were spores. Strains were not listed by authors in the literature.

[2]10^6 Organisms--Dose Range or "Survival-Kill Window" (i.e., some kill vs. all kill) exposed to a UV intensity of 400 uW/cm² for various periods of time.

[3] Represents % kill of 10^7 organisms per ml in water.

[4] Isolated from Alcon manufacturing area.

TABLE 7. UV resistance data for bacterial isolates

UV Resistance Category	UV Dose[1]	Bacterial Isolates[2]
Extremely Sensitive	< 1000 ergs/mm^2	19% (26/137)[3]
Sensitive	1000-5000 ergs/mm^2	46% (63/137)
Resistant	5000-50,000 ergs/mm^2	34% (47/137)
Extremely Resistant	> 50,000 ergs/mm^2	<1%[4] (1/137)

[1]UV dose obtained using UV intensity of 400 uW/cm^2 for various exposure times.

[2]Included vegetative cells and spores at a level of 10^4 to 10^5 organisms per inoculum spotted on the surfaces of agar plates (102 strains were environmental isolates and 35 strains were bioburden isolates).

[3]Number of strains surviving dose range/total number of strains tested.

[4]*B. pumilus* BMT-18 spores were the most resistant spores tested.

Ozone Sterilization in the Pharmaceuticals Industry

Gurley [13] believes that ozone generation and water/ozone contactor equipment now available make ozone a cost-competitive sterilization technique for the pharmaceuticals industry, especially in the area of parenterals. Ozone can be used gainfully in the preparation of sterile process water, for the terminal sterilization of aqueous drug formulations free from organic components, and in the sterilization of glass and certain plastic containers, notably polyvinyl chloride.

TABLE 8. Resistance data on selected organisms exposed to uW/cm^2 UV intensity

Organism	D-Value		
	Test #1	Test #2	Test #3
Micrococcus radiodurans ATCC 13939	33.25 (0.974)[1]	32.99 (0.988)	33.06 (0.981)
Staphylococcus aureus ATCC 6528	0.90 (0.988)	0.97 (0.986)	0.92 (0.986)
Streptococcus faecium ATCC 10541	2.00 (0.988)	1.97 (0.986)	2.02 (0.988)
Candida albicans ATCC 10231	7.31 (0.993)	7.51 (0.995)	7.55 (0.989)
Escherichia coli ATCC 8739	1.36 (0.983)	1.38 (0.989)	1.33 (0.998)
Pseudomonas aeruginosa ATCC 9027	0.63 (0.992)	0.57 (0.994)	0.60 (0.987)
Pseudomonas diminuta ATCC 19146	1.23 (0.984)	1.27 (0.981)	1.20 (0.983)
Pseudomonas diminuta ATCC 11568	1.94 (0.992)	2.02 (0.990)	1.97 (0.993)
Pseudomonas maltophilia ATCC 13637	1.14 (0.978)	1.17 (0.981)	1.19 (0.985)
Pseudomonas cepacia ATCC 25416	0.96 (0.988)	0.95 (0.982)	1.00 (0.987)
Pseudomonas putrefaciens ATCC 8071	1.48 (0.990)	1.41 (0.987)	1.44 (0.988)

[1]Correlation coefficient for each data set which represents the average of duplicate tests.

Empty glass bottles have been sterilized using ozone-containing air at concentrations of about 30 ppm of compressed gas. Control bottles inoculated with 10^3 yeasts, molds, spores, and vegetative cells were sterilized by the end of an 18-second period. The ozone-containing air was removed by sweeping with sterile air [34]. Schneider and Rump [33] used ozone-containing rinse solutions to sterilize glass containers. Control bottles contained 10^4 yeasts and bacterial cells. Ozone doses of 2.5 to 3.0 mg/l rendered them sterile upon rinsing. Pyrogenic material is destroyed rapidly by ozone. Glass and polyvinyl chloride containers are sterilized simultaneously and depyrogenated by ozone [19,27].

Gurley [13] advocates the use of ozone as a sterilant in the production of pharmaceutical process waters. His 1985 estimate of capital expenditure costs is in the $30,000 to $300,000 range, depending upon the size and complexity of the system.

Ozone in Integrated Circuit Waters

Ozone was utilized at the Bell Telephone Laboratories to eliminate bacteria from ultrapure rinsewater [38]. Ozone was fed into a storage tank over a fifteen minute interval. The ozonized water was then permitted to stand several hours until the ozone vanished by decomposition. The practice was repeated twice weekly to yield, 5 to 20 bacteria per ml and on occasion, undetectable bacteria levels.

Ozone and Hydrogen Peroxide

Hango *et al*. [14] utilized ozone in combination with hydrogen peroxide in the reclamation of recycled semiconductor rinsewater. The removal of organic compounds, largely isopropanol and acetone, was sought. At pH 7.5 and using a weight ratio of 5:1 ozone to hydrogen peroxide, TOC levels of 18 mg/l were completely removed in 60 minutes. The ratio of ozone to TOC was 14:1. Neither ozone by itself, nor with ultraviolet radiation, was as cost-effective. The use of 1.7 lbs of ozone plus 0.42 lb of H_2O_2 per 1,000 gal was estimated at 1981 prices to cost 83 cents per 1,000 gal.

UV Radiation and Ozone in Water Systems

Francis [9] prefers the combination of UV radiation and ozone for the removal of TOC. Ultraviolet radiation by itself does not occasion the total oxidation of organic materials. Also, its microbial kill rate, particularly of spores, is lower than that of ozone. Ozone itself reacts more slowly with organics than does the hydroxyl free radical, and is less effective against the carboxylic acids than are some of the standard intermediates produced during the oxidation of organic compounds. Ozone transformed into hydroxyl free radicals by combination with UV radiation of 254 nm wavelength is the course favored by Francis.

The combination of ozone and UV offers another advantage in addition to the conversion of ozone into a more aggressive oxidizer. The solubility of ozone in water is relatively low (about 13 times higher than oxygen, depending upon temperature and concentration of ozone in the gas phase), and its transference from the gas phase into the water phase reflects, at any partial pressure, its concentration gradient between gas and water. The ozone-decomposing action of the UV light serves to enhance the gas to liquid mass transfer of the ozone by converting it to hydroxyl free radicals in the water.

In one study, water was circulated through an ozone/UV reactor at a flow rate of 240 l/h. A steady state TOC level of 100 μg was reduced to under 20 μg. Series or parallel arrangements of such reactors could be used to achieve lower TOC levels at fixed flow rates, or to obtain such TOC values at higher flow rates. This technique lends itself to the recovery of rinsewater, its customary application in the semiconductor industry, as well as the primary preparation of low TOC water. In either case, D.I. resins are utilized for the removal of the carboxylic acids that are generated oxidatively. The introduction of ozone or hydroxyl radicals into D.I. beds can result in the degradation of the resins with concomitant generation of TOC.

The current interest in ozone in the United States is its use for the inactivation of Giardia and enteric viruses in the treatment of drinking water. The EPA's soon-to-be-proposed surface water treatment regulations conclude that ozone is the most effective disinfectant, and encourage its adoption as the primary disinfectant. Combinations of ozone with UV radiation or with hydrogen peroxide show great promise for the oxidative destruction of organic contaminants in potable water supplies, particularly those pollutants that are most difficult to remove by current techniques.

Safety Issues and Provisions with Respect to Ozone and UV Radiation

Because ozone is one of the most powerful oxidizing agents known to man (see Table 1), it should be considered as hazardous material, and one which should be handled carefully. In practice, well designed ozonation systems incorporate subsystems for the destruction of excess ozone which may be present after the ozone contact chamber. Such subsystems insure that no ozone will leak into the ambient atmosphere during normal operations.

To provide for accidental discharges of ozone, however, it is customary to install an ambient air ozone monitor at an appropriate point or points in the equipment room(s). These monitors are usually designed to signal an alarm at an ozone level of 0.1 ppm. Additionally, such monitors can be interfaced with exhaust fans as well as electrical lines to the ozone generator itself. Once electricity to the ozone generator is cut, ozone generation ceases immediately. Exhaust fans rapidly remove any leaked ozone from the room, preventing large quantities of ozone from accumulating. The ozone ambient air action level of 0.1 ppm has been adopted by the Occupational Safety & Health Administration as its Maximum Exposure Level for exposure of workers to ozone on a time-weighted average over an 8-hour working day, five days per week.

UV radiation is also hazardous, in the sense that it can damage the human cornea. Consequently, UV-producing bulbs are normally contained in housings which prevent UV radiation from exiting the equipment. In addition, many UV systems are designed so that whenever the housing to the UV bulb is opened inadvertently during operation, the electrical supply to the equipment is stopped, thus eliminating the possibility of inadvertent exposure to UV radiation.

As is the case with the commercial use of many hazardous materials, understanding the properties and potential hazards associated with ozone and with UV radiation is fundamental to the design and operation of the appropriate equipment.

Acknowledgments

We would like to acknowledge Greg E. Ellner of Continental Water (San Antonio, TX) for his assistance in preparing this chapter.

References

1. Abshire, R.L. 1986. The use of UV radiation as a method of sterilization in the pharmaceutical industry. p. D.1.1-D.1.19. In W.J. Maschelein (ed.), Proceedings of the International Ozone Association, European committee, Paris, 1986.

2. Abshire, R.L., B.L. Schlech, and H. Dunton. 1983. The development of a biological indicator for validating ultraviolet radiation sterilization of polyethylene bottles. J. Parent. Sci. and Technol. 375:191-197.

3. Antopol, S.C. and P.D. Ellner. 1979. Susceptibility of *Legionella pneumophilia* to ultraviolet radiation. Appl. Environ. Microbiol. 38:347-348.

4. Blogoslawski, W.J., C. Brown, E.W. Rhodes, and M. Broadhurst. 1975. Ozone disinfection of a seawater supply system. p. 674-687. In R.G. Rice and M.E. Brown (ed.), Proceedings of the 1st international symposium on ozone for water and wastewater treatment. International Ozone Association, Norwalk, CT.

5. Coin, L., C. Gomella, C. Hannoun, and J.-C. Trimoreau. 1967. Ozone inactivation of poliomyelitis virus in water. La Presse Medicale 72(37):1883-1884.

6. Coin, L., C. Hannoun, and C. Gomella. 1964. Inactivation of poliomyelitis virus in the presence of water. La Presse Medicale 72(37):2153-2156.

7. Dobiasch, P.H.S. 1986. Principles of combined U.V. and ozone generation. p. B.3.1-B.3.7. In W.J. Masschelein (ed.), Proceedings of the

International Ozone Association, European committee, Paris.

8. Ellner, P.D. and S. Ellner. 1986. A biological method for measuring the actual dosage delivered by ultraviolet water purifiers. Private Communication.

9. Francis, P.D. 1987. The use of ultraviolet light and ozone to remove organic contaminants in ultrapure water. Proceedings of the Interated Circuit Manfacturers Consortium Conference, Feb. 9-10, Coventry, U.K.

10. Gelzhauser, P. 1986. Application and design of UV units. p. B.2.1-B.2.19. In W.J. Masschelein (ed.), Proceedings of the International Ozone Association, European Committee, Paris.

11. Glaze, W.H. 1987. Drinking water treatment with ozone. Environ. Sci. Technol. 21(3):224-230.

12. Gordon, G., and G.E. Pacey. 1986. An introduction to the chemical reactions of ozone pertinent to its analysis. p. 41-52. In R.G. Rice, L.J. Bolyky, and W.J. Lacy (ed.), Analytical aspects of ozone treatment of water and wastewater treatment, Lewis Publishers, Inc., Chelsea, MI.

13. Gurley, B. 1985. Ozone: pharmaceutical sterilant of the future. J. Parent. Sci. Technol. 39(6):256-261.

14. Hango, R., F. Doane, and L.J. Bollyky. 1981. Wastewater treatment for reuse in integated circuit manufacturing. p. 303-313. In O.H. Hess (ed.), Colloquium Verlag Wasser Berlin, FRG.

15. Harm, W. 1980. Biological effects of ultraviolet radiation. Cambridge University Press, Cambridge, U.K.

16. Joel, A.R. 1987. Sorbozon: Ozone Generation with internal oxygen recycle. p. 83-88. In D.W. Smith and G.R. Finch (ed.), The role of ozone in water and wastewater treatment, Kitchener, Ontario, Canada.

17. Majumdar, S.B.W. Cechler, and O.J. Sproul. 1974. Inactivation of poliovirus in water by ozonation. J. Wat. Pollut. Cont. Fed. 46:2048-2053.

18. Majumdar, G.B. and S.J. Glassman. 1974. A viable treatment for circulating cooling water systems. Paper presented at the 36th annual American power conference, IL.

19. Masaoka, T., Y. Kubota, S. Namiuchi, T. Takubo, T. Ueda, H. Shibata, H. Nakamura, J. Yoshitake, T. Yamayoshi, H. Doi, and T. Kamiki. 1982. Ozone decontamination of bioclean rooms. Appl. Environ. Microbiol. 43:509-513.

20. Meltzer, T. 1987. Water purification. p. 791-918. In T. Meltzer (ed.), Filtration in the pharmaceutical industry, Marcel Dekker, NY.

21. Merrill, D.T. and J.A. Drago. 1980. Evaluation of ozone treatment in air-conditioning cooling towers. p. 487-495. In J.F. Garey, R.M. Jorden, A.H. Aitken, D.T. Burton, and R.H. Gray (ed.), Condenser biofouling control, Ann Arbor Science, Ann Arbor, MI

22. Meulemans, C.C.E. 1986. The basic principles of uv-sterilization of water. p. B.1.1-B.1.13. In

W.J. Maschelein (ed.), Proceedings of the International Ozone Association, European committee, Paris.

23. Morris, J.C. 1975. Aspects of the quantitative assessment of germicidal efficiency. p. 1-10. In J.D. Johnson (ed.), Disinfection: Water and wastewater, Ann Arbor Science, Ann Arbor, MI.

24. Nakayama, S., M. Tanaka, S. Yamauchi, and N. Tabata. 1980. An anti-biofouling ozone system for cooling water circuits. I. Application to fresh water circuits. Ozone: Sci. Engin. 2(4):327-336.

25. Nakayama, S., M. Tanaka, S. Yamauchi, and N. Tabata. 1985. Anti-biofouling ozone system for cooling water circuits. II. An application to seawater. Ozone: Sci. Engin. 7(1):31-45.

26. Nebel, C. 1981. Ozone. p. 683-713. In K. Othmer (ed.), Encyclopedia of chemical technology, Vol. 16, John Wiley, NY.

27. Nebel, C., and T. Nebel. 1984. Ozone: the process water sterilant. Pharm. Manufact. 1(2):16-22.

28. Nebel, C. and P. Unangst. 1973. An evaluation of various mixing devices for dispensing ozone in water. Wat. Sewage Works, ref. no. 1973, p. R-6.

29. Nebel, C., R.D. Gottschling, R.L. Hutchinson, T.J. McBride, D.M. Taylor, J.L Pavoni, M.E. Tittlebaum, H.E. Spencer, and M. Fleischman. 1973. Ozone disinfection of industrial-municipal secondary efluents. J. Wat. Pollut. Cont. Fed. 45:2493-2507.

30. Rice, R.G. and L.J. Bollyky. 1981. Fundamental aspects of ozone technology for cooling tower water treatment. p. 1-19. In R.G. Rice (ed.), Ozone treatment of water for cooling applications, International Ozone Association, Norwalk, CT.

31. Rice, R.G. and J.C. Hoff. 1981. Inactivation of Giardia lamblia cysts by ultraviolet radiation. Appl. Environ. Microbiol. 42:546.

32. Rice, R.G., L.J. Bolyky, and W.J. Lacy. 1986. Analytical aspects of ozone treatment of water and wastewater, Lewis Publishers, Chelsea, MI.

33. Schneider, W.K.G. and H.H. Rump. 1983. Use of ozone in the technology of bottled water. Ozone: Sci. Engin. 5:95-101.

34. Stopka, K. 1981. Our experiences with 14 ozone-treated cooling towers. p. 55-61. In R.G. Rice (ed.), Ozone treatment of water for cooling applications, International Ozone Association, Norwalk, CT.

35. Sugam, R. and C.R. Guera. 1981. Comparison of chlorine and ozone for power plant cooling water treatment. p. 63-73. In R.G. Rice (ed.), Ozone treatment of water for cooling applications, International Ozone Association, Norwalk, CT.

36. Toriceli, A. 1959. Sterilization of empty containers for the food industry. p. 375-380. In Advances in chemistry series, Vol.21, Ozone chemistry and technology, American Chemical Society, Washington, D.C.

37. Wickramanayake, G.B., A.J. Rubin, and O.J. Sproul. 1985. Effects of ozone and storage tem-

perature on *Giardia* cysts. J. Am. Wat. Works Assoc. 77(8):74-77.

38. Zmolek, C.R. 1977. Ultrapure water for integrated circuits processing. Indust. Wat. Engin. Dec. p. 6-11.

CHAPTER 6. MICROBIAL FOULING OF REVERSE OSMOSIS MEMBRANES: GENESIS AND CONTROL

Harry F. Ridgway, Orange County Water District, Fountain Valley, CA

Background

Reverse osmosis (RO) is a cross-flow, membrane separation process commonly employed for the demineralization (or decontamination) of aqueous process streams. The essential feature of all RO systems is a semipermeable membrane, which preferentially allows the passage of water molecules while simultaneously impeding the transport of most organic [68] and inorganic solutes [27,36,41,58,65]. Since RO membranes do not possess a visible pore structure, even when they are viewed by electron microscopy [36,41, 58,65], they also completely exclude (i.e., reject) virtually all suspended particulate material and colloidal substances, including bacteria and viruses [36, 58,69].

Nearly thirty years ago, Loeb and Sourirajan [47] described the first practical, asymmetric, cellulose acetate RO membrane. A significant improvement in membrane separation technology came much later with the introduction of the first polyamide, thin-film-composite membrane in the early 1970's [10,65,77]. Since that time, there has been a continual evolution and proliferation of new membrane polymers and module designs for a wide diversity of commercial applications [58,61,65,77].

Because of its efficiency, reliability, and overall cost-effectiveness, demineralization by RO has become a central and indispensable component for a wide variety of industrial and municipal applications. Among the primary uses of RO are (1) the production of ultrapure water for the pharmaceutical and electronics (semiconductor) manufacturing industries [58,65, 82,88], (2) advanced wastewater treatment and reuse applications [1,3,4,40,56,90,93,94], and (3) brackish or seawater demineralization for the production of potable water [21,27,51,83,89]. In addition, RO demineralization is now also widely utilized in clinical laboratory applications [65], in medicinal applications such as artificial kidney dialysis [36, 41,82,88], in the separation and purification of the products of biotechnology [36,58,88], and in the food industry (e.g., milk and whey processing)[36,41, 87].

Chemical and biological fouling of the individual structural components which comprise the RO module can sharply reduce the cost-efficiency of the RO process (Figure 1). The semipermeable membrane surfaces in RO systems are particularly susceptible to microbial colonization and biofilm formation. This is due to the extremely large areas of membrane surface that are continually exposed to contaminated feed streams [1-4,36,37,42,43,51,54,69-71,73-75,94]. When biofouling of the RO membrane surface occurs, it can result in a dramatic decline in overall membrane performance (i.e., a reduction in water permeation and mineral rejection). The primary purpose of this chapter will be to provide a broad overview of the underlying causes and effects of biological fouling in RO membrane systems. Particular attention will be paid to the role that the bacterial adhesion process plays in biofilm genesis, and how this process adversely affects the performance characteristics of the semipermeable membranes. In addition, some strategies and methods for the control of biological fouling in RO systems will be examined.

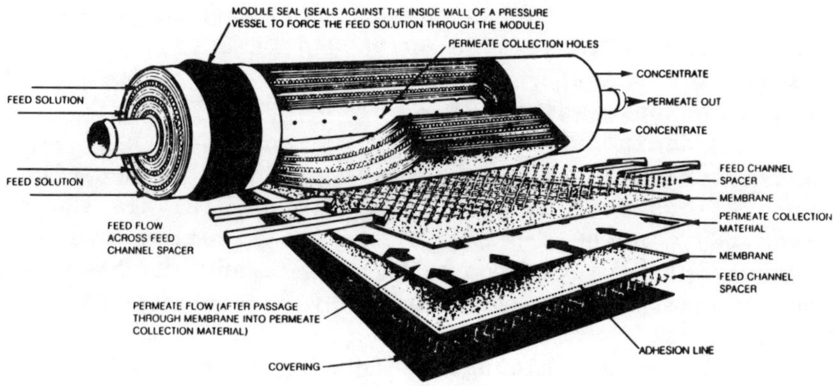

FIGURE 1. Diagrammatic illustration of typical spiral-wound RO module. Note tangential nature of feedwater flow over membrane surfaces. From Ridgway _et al_. [71].

Biofilm Genesis in RO Systems

One of the principle causes of RO membrane performance decline is the formation of an attached microbial biofilm on the feedwater (i.e., semi-permeable) surface of the membrane. Not only does the development of a biofilm result in a significant reduction in the net permeability of the RO membrane to water (i.e., a reduction in membrane flux), but also the differential operating pressure (i.e., Δp) of the system will tend to increase over time. In addition, the presence of a biofilm is frequently associated with a gradual loss of the mineral rejection characteristics of the membrane. These and other relevant aspects of RO biofouling will be discussed in more detail in a subsequent section of this chapter.

Microbial Adhesion

Before a biofouling layer can develop on the RO membrane surface, viable microorganisms which are characteristically transported in the feedwater must first contact and then adhere to the membrane. Once a bacterium enters the RO module in the feedwater, a variety of factors can then influence its ability to establish intimate contact with the membrane surface. Among the more significant factors that can promote contact of the bacterium with the RO membrane surface are (1) convective or turbulent flow near the membrane surface; (2) thermally-induced (i.e., Brownian) motion; (3) sedimentation under the influence of gravity, which is especially relevant when the RO system is quiescent; and (4) motility and chemotaxis exhibited by many fouling microorganisms. References 15 and 49 contain a more elaborate discussion of each of these factors as they relate to microbial surface adhesion processes in general. In RO systems, elevated operating pressures and system (permeate) recoveries are also known to promote initial contact of bacteria (or other suspended particles) with the membrane surface [40,64,66,90]. However, because 85-90% of the feedwater flow in a typical RO system is in a direction which is tangential relative to the membrane surface [36,58,89], bacteria and other colloidal particles should, theoretically, be continually swept away from the RO membrane surface into the bulk fluid phase (i.e., into the concentrate or brine stream). Yet, in actual practice such sweeping action does not occur with absolute efficiency and many bacteria, when given sufficient residence time in close proximity to the membrane surface, will irreversibly attach (Figure 2). Thus, it is the nature of the adhesion interaction which occurs between specific chemical ligands that are exposed on the bacterial cell surface and the RO membrane surface that will, to a large extent, determine the rate and measure of nascent biofilm formation. In the absence of such mutual attractive forces, bacteria and other particles could be expected to be continually removed from the RO membrane surface by fluid shear forces, thereby precluding the development of an attached biofilm.

The mechanism of bacterial adhesion to cellulose acetate RO membrane surfaces was recently investigated by Ridgway and coworkers [73-76] using an in vitro bacterial adhesion assay technique. Radioisotopically-labeled Mycobacterium BT2-4 cells (Figure 3) were utilized in the adhesion studies. This microbe was previously found to be involved in the earliest stages of RO biofouling at Water Factory 21, a 15 mgd, prototype advanced wastewater reclamation facility where the investigation was conducted [70,73]. A suspension of the uniformly-radiolabeled mycobacteria was placed in a sterile, plastic, screw-cap test tube. The bottoms of the test tubes had been previously cut off so that it was necessary to first cap the tube and then invert it before adding the bacterial suspension through the open end; however, prior to screwing the cap into position, a circular disc of the RO membrane to be tested was used to replace the cap liner. The membrane disc was oriented such that the active, semipermeable surface was in direct contact with the bacterial suspension.

Following a period of incubation under standard conditions of temperature and mixing velocity, non-adherent bacteria were rinsed off the surface of the RO membrane disc, and the number of firmly (i.e., irreversibly) attached mycobacteria were quantified by measuring the total amount of membrane-bound radioactivity in a liquid scintillation counter. Since no hydraulic pressure differential was present across the RO membrane discs, only the affinity of the bacterial cell for the RO membrane surface was measured in such experiments. The adhesion of the mycobacteria to the cellulose acetate membrane discs was surprisingly rapid, with no apparent lag phase [73,76,78].

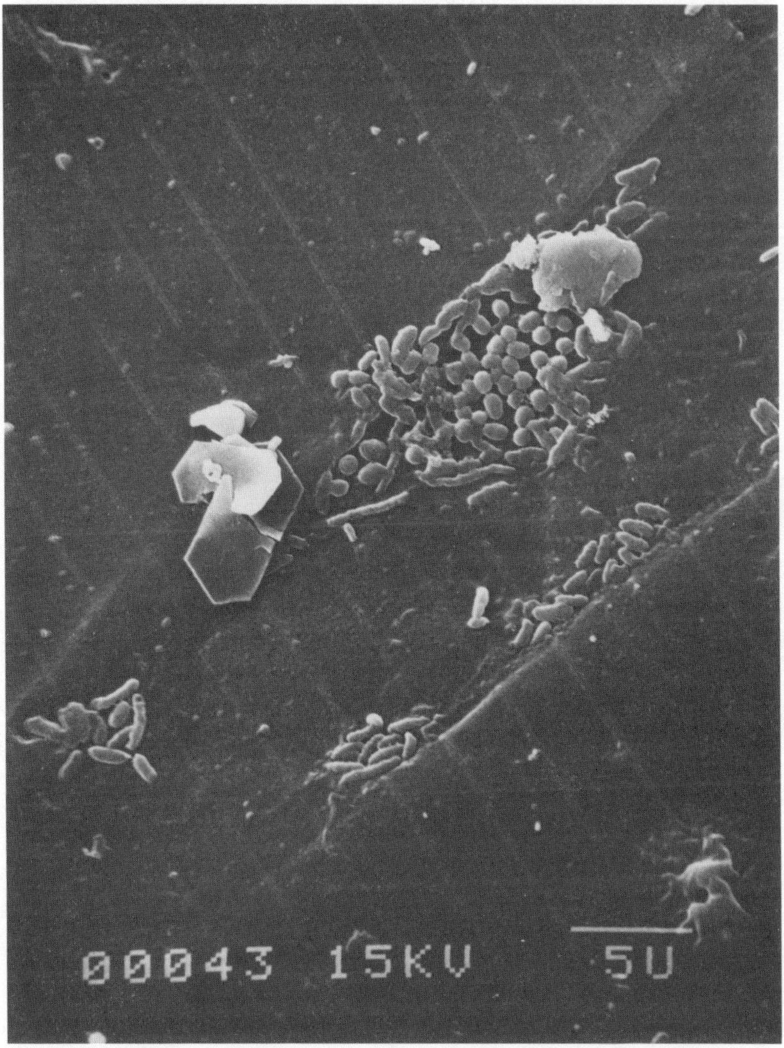

FIGURE 2. SEM photo of fouling bacteria recently attached to a cellulose acetate RO membrane surface. Note that these early microcolonies have begun to form in a shallow depression in the membrane surface where turbulence tends to be minimized. Bar = 5.0 μm.

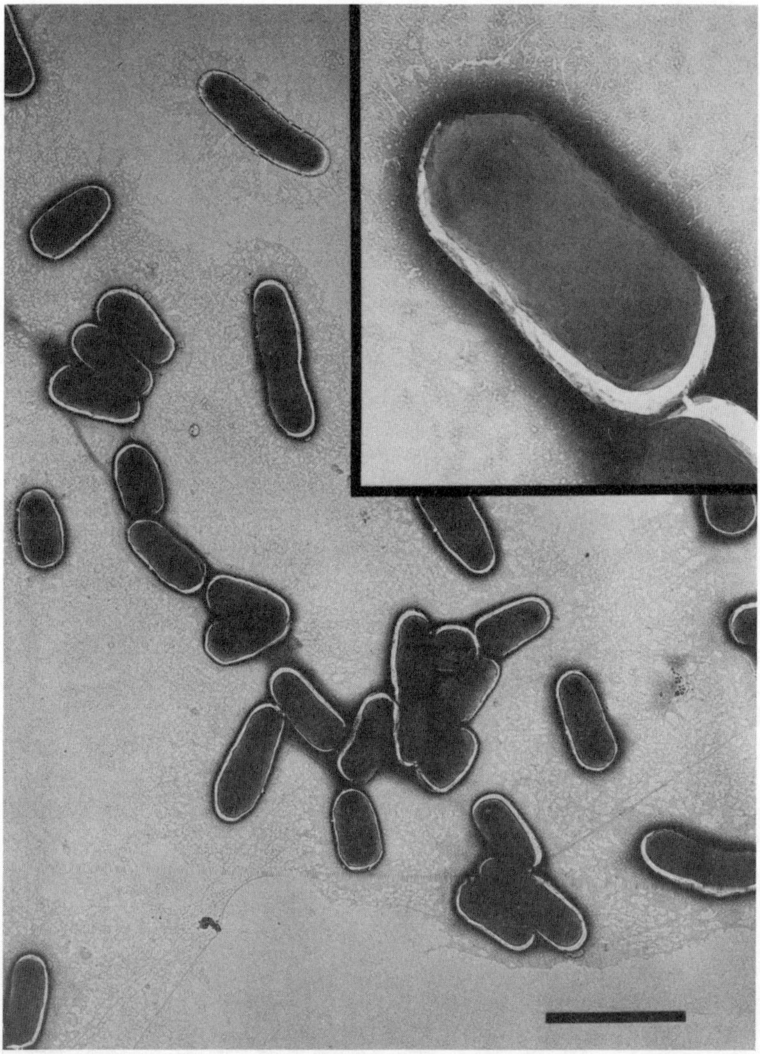

FIGURE 3. TEM preparation of Mycobacterium BT2-4 cells negatively stained with potassium phosphotungstic acid. Note fibrillar glycocalyx material surrounding cells which is thought to enhance their ability to adhere to RO membrane surfaces. Bar = 1.0 μm.

Thus, once cells contact the RO membrane surface, they begin to attach immediately. Although the fluid velocity and overall hydrodynamics could be expected to differ considerably, microorganisms in an actual RO system would presumably respond similarly in terms of their adhesive behavior. The kinetics of mycobacterial adhesion in the laboratory assay were biphasic: an initial rapid phase of adhesion, which was more-or-less linear with time, followed by a much slower, also nearly linear rate of attachment (Figure 4). The initial, rapid adhesion phase typically lasted for

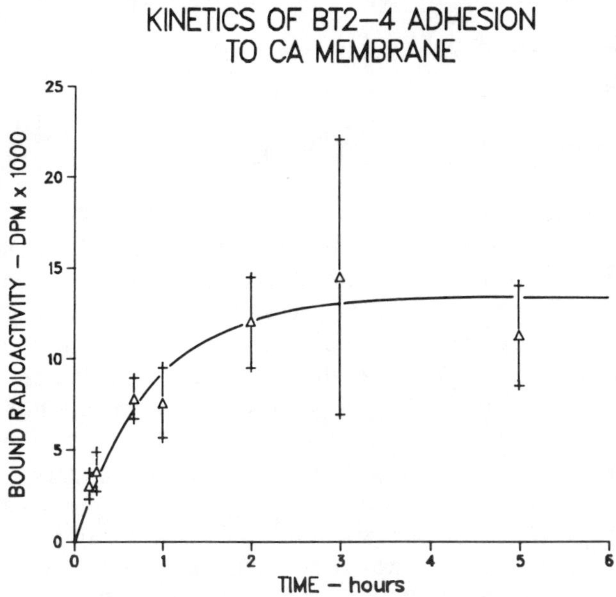

FIGURE 4. Kinetics of mycobacterial adhesion to cellulose acetate RO membranes as determined using _in vitro_ laboratory test procedure described by Ridgway _et al_. [76]. Adhesion of the radiolabeled bacteria is indicated by the amount of bound radioactivity (disintegrations per minute; DPM) associated with the RO membrane. Bacteria added at time 0 = 2.87×10^7 per ml.

1-2 hours, while the slow adhesion phase continued indefinitely, or until termination of the experiment. It was demonstrated in the same study that the mycobacteria exhibited typical saturation adsorption kinetics that could be described by the Langmuir adsorption isotherm equation [44]. Thus, the total amount of bacterial adhesion observed was directly proportional to the starting (i.e., free) cell number; however, when the point of saturation binding was attained, no further adhesion occurred, even if the free bacterial concentration at the beginning of the assay was increased (Figure 5). This type of surface absorption behavior is not unlike the adhesion kinetics that have been reported for many other bacterial types [6,49].

The foregoing observations are important in terms of biological fouling of RO membrane surfaces, since they imply that a higher feedwater concentration of microorganisms will lead to a proportionately increased rate of membrane biofouling. This finding tends to substantiate the need for pretreatment processes, such as submicron or ultrafiltration, which are capable of physically removing microorganisms from the feedwater stream. The data further indicate that the RO membrane surface possesses a finite number of available bacterial adsorption sites. Once the putative adsorption sites are fully occupied by the fouling bacteria, no additional adhesion may take place. The total number of mycobacterial adsorption sites on the cellulose acetete membranes studied was estimated to be about 3.0×10^6 per cm^2, which was approximately one order of magnitude less than the theoretical binding capacity of the membranes based solely upon spacial and geometric considerations [73]. No information is currently available concerning the physico-chemical nature, distribution, or specificity of the adsorption sites.

Among the primary forces that have been demonstrated to be involved in the adhesion of bacteria and other microorganisms to submerged or wetted solid surfaces are long-range electrostatic and van der Waals interactions [49,55], as well as short-range ionic [30,31,60], hydrogen bonding [49,92], and hydrophobic interactions [22,29,30,48,49,60,74,75]. Hydrophobic interactions have been shown in recent years to be of particular significance in the adhesion of many different kinds of bacteria to a wide variety

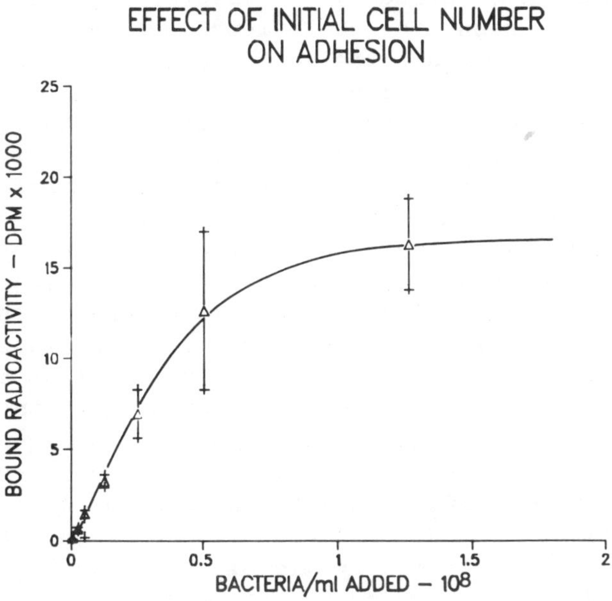

FIGURE 5. Mycobacterial adhesion to cellulose acetate RO membrances as a function of applied bacterial concentration. Adhesion of radiolabeled cells is indicated by amount of radioactivity (as disintegrations per minute; DPM) bound to RO membrance. Note that adhesion sites on the RO membrance surface become fully saturated when approximately 1 x 10^8 cells per ml are added. Adhesion time was five hours.

of natural and synthetic substrate types [22,29,30, 48,49,60]. The involvement of hydrophobic interactions in the adhesion of mycobacteria to cellulose diacetate RO membranes was recently experimentally demonstrated by Ridgway _et_ _al_. [74,75]. Using a modification of the adhesion assay described above, it was found that dramatic fluctuations in the ionic environment of the cells, such as large changes in the net ionic strength of the suspending medium, or the addition of high molecular-weight charged polymers, were markedly less effective at interfering with mycobacterial attachment to the RO membranes than was the addition of a low concentration of a nonionic surfactant, such as Triton X-100.

Furthermore, a direct correlation was observed between the relative hydrophobicity of a bacterial cell and its ability to adhere to the RO membrane surface. Such a result may not be too surprising given the extremely hydrophobic cell exterior which characterizes virtually all mycobacteria [24,52,74,75], as well as the partially hydrophobic nature of the RO membrane itself [41,65]. Surface active agents are able to disrupt mutual hydrophobic interactions between such surfaces, presumably by binding to and effectively masking the exposed hydrophobic ligands [38], leaving the hydrophilic moiety of the bound detergent free to interact with the aqueous environment. This concept of surfactant interaction, as it pertains to RO specifically, is diagrammatically illustrated in Figure 6, and represents a compelling justification for the incorporation of detergents in many RO cleaning solutions. Further discussion of the use of surfactants in membrane cleaning formulations may be found in the section on Control of Membrane Biofouling.

Depending upon the specific chemical environment proximal to a submerged solid surface, long-range electrostatic interactions may also contribute to the adhesion of fouling microorganisms [30,49,60]. For example, the rate and extent of adhesion of mycobacteria to cellulose acetate type RO membranes was shown to be greatly enhanced, sometimes by more than ten-fold (compared to an untreated control preparation), by the addition of certain cationic quaternary ammonium surfactants to the bacterial suspension [76]. This "adhesion enhancement" effect was observed despite the fact that the cells had been metabolically inactivated by the surfactant [78]. This suggests that initial attachment may be governed principally by non-physiological interactions. Such a conclusion is contradictory to results recently reported by Paul [59] and Salit [79], which suggest that several metabolic processes, including cellular energy transduction and macromolecular biosynthesis may be required for stable adhesion to occur. The stimulation of mycobacterial adhesion by the quaternary amines was strongly concentration dependent, which presumably resulted from the differential binding of detergent molecules to both the bacterial cell surface as well as to the RO membrane.

Within a narrow range of quaternary amine concentration (0.01% to 0.2% wt./vol.), the charge differential between the cell surface and the RO membrane surface is evidently maximized, due to preferential adsorption of the surfactant to the bacterial cell surface components. The strong, net cationic charge imparted to the cell by the bound surfactant molecules then mediates adhesion of the bacteria to the more electronegative RO membrane surface. Additional evidence was presented in the same investigation [76] which indicated that a clear structure-function relationship exists between the molecular configuration of a quaternary ammonium compound and its ability to elicit the adhesion enhancement effect. A dodecyl alkyl group was found to be an essential component of detergent structure for the adhesion enhancement effect to be observed. This suggests the possibility that the length of the alkyl chain may be important in determining the stability of interaction between the quaternary amine and the mycobacterial cell envelope [76,78]. An analogous structure-function relationship has been documented for the antibacterial activi-

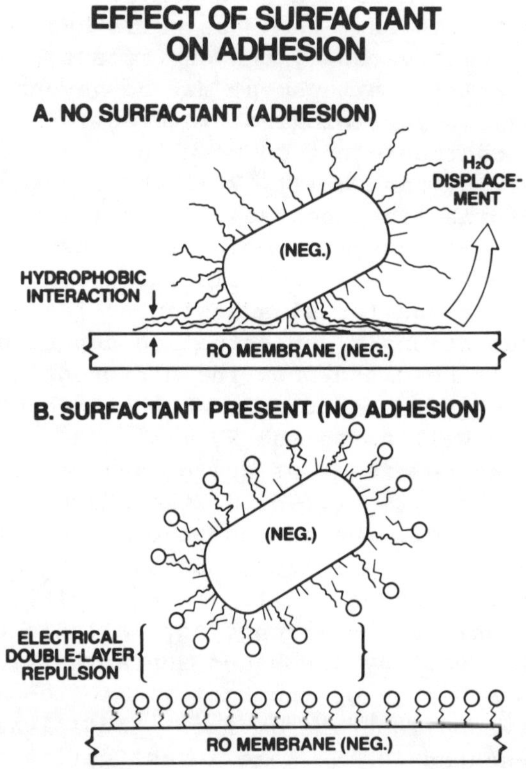

FIGURE 6. Diagrammatic illustration of surfactant activity. Displacement of a structured layer of absorbed water by approaching surfaces increases system entropy, which drives the reaction forward (upper panel). In the presence of a surfactant, hydrophobic sites on the cell and RO membrane are masked by detergent molecules, thereby precluding a hydrophobic interaction (lower panel). From Ridgway _et_ _al_. [74].

ties of a homologous series of quaternary ammonium compounds possessing different length alkyl chain lengths [62].

Further evidence for the involvement of strong electrochemical attractions in the adhesion of fouling microorganisms to RO membrane surfaces has emerged from an examination of the kinetics of mycobacterial adhesion to a variety of RO membrane types differing in their surface charge characteristics [Ridgway, unpublished data]. Interestingly, the greatest degree of mycobacterial adhesion has been observed using certain polyamide type RO membranes, which possess a strongly anionic nature due to free carboxyl or sulfonate groups. Less strongly charged membranes, such as cellulose acetate, generally exhibited a five to ten-fold reduction in bacterial adhesion compared to the more highly charged membranes.

The experimental findings described above suggest that both hydrophobic and electrochemical attractive forces are involved in the adhesion of fouling bacteria to RO membrane surfaces. Which types of attractive forces predominate appears to be dependent upon a number of environmental variables, including the specific ionic conditions, the chemistry and structure of surfactants which may be present, the type of RO membrane, and the type of fouling bacterium. In the absence of more detailed information concerning each of these parameters, it is impossible at this time to extrapolate with confidence from the results of the laboratory adhesion experiments to actual field operating experiences. Nevertheless, the laboratory data are extremely valuable, since they have added to our understanding of the fundamental nature of the bacterial adhesion mechanism, which is necessary before rational strategies can be developed for the control of biofouling in RO systems.

Kinetics of Biofilm Formation

The rate at which a surface biofilm develops is determined by the interaction of various physicochemical, hydrodynamic, and biological factors. It is obvious from the foregoing discussion that the kinetics of biofilm formation on RO membranes will depend partly on the concentrations and specific adhesion properties of the individual fouling microorganisms and partly on their growth rates once they have attached to the membrane surface. Characklis and coworkers [13-15], as well as numerous other investigators [8,18,50], have quantified some of the more significant nutritional and fluid dynamic interactions that regulate biofouling kinetics on submerged metallic or glass surfaces. Besides fluid shear and turbulent flow characteristics in proximity to the membrane surface, the membrane flux, various operating parameters (e.g., percent recovery, operating pressure, etc.), RO module configuration, and pretreatment level must also be considered in any analysis of biofouling rates in RO systems.

Very few studies on the kinetics of biofouling in RO systems have been conducted. Ridgway and coworkers [70] studied the rate of biofilm formation in small (two and one-half inch diameter), spiral-wound RO modules which were operated for various periods of time on an extensively pretreated municipal wastewater effluent. In spite of a low combined-chlorine residual (about 1-3 mg per liter as monochloramine), the total number of viable bacteria in the pretreated feedwater stream ranged from 1 x 10^3 to nearly 5 x 10^4 per ml [measured as colony-forming-units (CFU's)] for the duration of the study. Surprisingly, bacteria colonized the cellulose acetate RO membrane surfaces rapidly with approximately 3 x 10^5 CFU's per cm^2 being recovered after only three days of operation. In a similar study, the total number of attached bacteria, based on direct epifluorescent microscopic counts, was approximately 10^7 per cm^2 (Figure 7). Viable counts were considerably lower, due to a higher combined chlorine residual in the feedwater. This rapid colonization was observed after only three days of operation. Such rapid colonization by microorganisms was confirmed by direct visualization of the RO membrane surfaces by scanning electron microscopy (Figure 8).

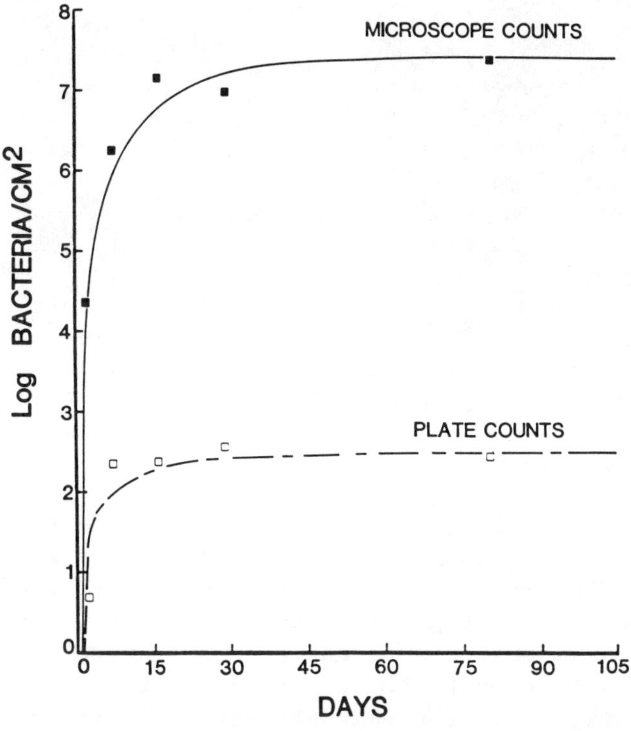

FIGURE 7. Rate of biofilm formation on RO membranes operated on pretreated municipal wastewater. Due to combined chlorine residual in feedwater, most of the attached bacteria are dead. Total bacteria (viable and nonviable) were enumerated by epifluorescent microscopy. Viable bacteria were determined by plate counts (open symbols).

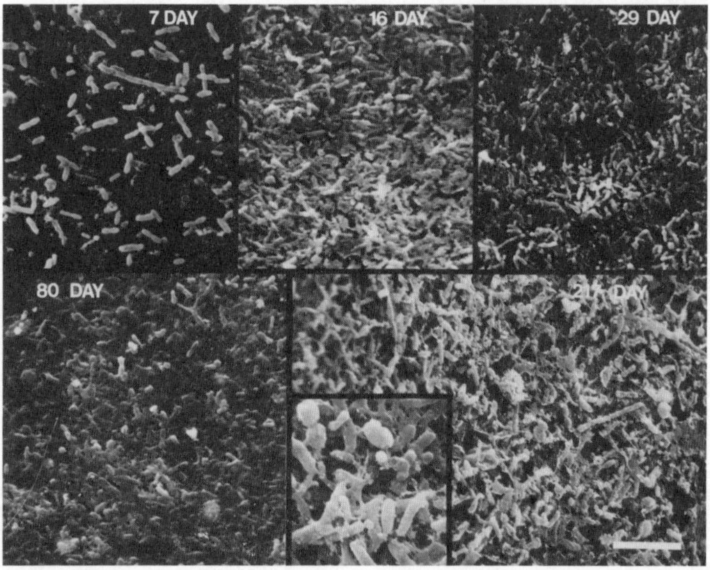

FIGURE 8. Series of SEM photos depicting RO biofilm development. Note glycocalyx material associated with attached cells. Bar = 5.0 μm.

The accumulation of protein, carbohydrate, and muramic acid (an amino sugar associated only with the bacterial cell wall) provided further evidence of rapid biofilm development. The deposition of such biologically-derived substances on the RO membrane surfaces was shown to be inversely correlated with the membrane flux. It was demonstrated in the same study that the continuous addition of 10 mg per liter of monochloramine to the RO feedwater completely inactivated the attached fouling bacteria, although it failed to prevent their adhesion to and accumulation on the membrane surfaces. Nor was the flux decline prevented by the continuous chlorine treatment.

Thus, the early stages of biofilm formation appear to be regulated primarily by non-metabolic, physico-chemical processes; namely, the affinity of the fouling bacteria for the RO membrane surface. Such a conclusion is consonant with the adhesion enhancement effect induced by certain quaternary ammonium surfactant/biocides, which was discussed earlier. A summary of the overall time course of events that are believed to punctuate the development of a biofilm on an RO membrane surface is provided in Figure 9.

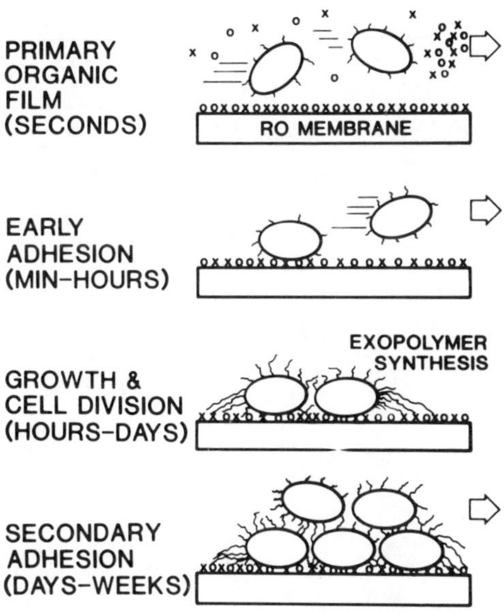

FIGURE 9. Diagram illustrating general course of events during RO biofouling. The approximate time frame for each event is also indicated.

Fouling Microorganisms

The identities of fouling bacteria isolated from various stages of biofilm development were also determined in the system described above. Curiously, only bacteria belonging to the acid-fast genus Mycobacterium were recovered from the RO membrane surfaces during the initial 43 days of operation [70]--despite the presence of large numbers of other microbial types in the RO feedwater (Table 1).

TABLE 1. Identification of bacteria from RO membranes[1]

	Days of Operation						
Taxonomic Group	3	5	11	20	43	57	215
Mycobacterium	100	100	100	100	100	93	0
Acinetobacter	0	0	0	0	0	0	52.9
Shigella	0	0	0	0	0	0	2.0
Alcaligenes	0	0	0	0	0	0	5.9
Pseudomonas	0	0	0	0	0	0	23.5
Klebsiella	0	0	0	0	0	0	2.0
Flavobacterium/ Moraxella	0	0	0	0	0	0	2.0
Gram-positive	0	0	0	0	0	2.3	9.8
Unidentified	0	0	0	0	0	4.7	2.0
Number of isolates tested	40	40	40	40	40	43	51

[1]Column numbers indicate percent of isolates identified for each membrane. Data from Ridgway et al. [70].

Following colonization by the pioneering population, a variety of other bacterial genera became increasingly prevalent on the membrane surface. After 215 days of operation, the mycobacteria represented less than two percent of the sessile population, while *Pseudomonas* and *Acinetobacter* accounted for approximately 53 and 24 percent, respectively, of the same population. The latter two microbial genera, together with *Flavobacterium* spp., were also found to predominate on cellulose acetate RO membranes that had been in operation for periods of up to three years [70,71]. These, and similar observations [Ridgway, unpublished data] indicate that the mycobacteria play an important role in the early genesis of the biofilm, and that they may be instrumental in the initiation of a true microbial succession on the RO membrane surface. Unfortunately, it is not known at this time whether such findings may be generalized to include other RO facilities.

It should be noted that many of the microorganisms that have been recovered from the biofouled RO membrane surfaces, including the mycobacteria, are considered to be opportunistic pathogens capable of producing infection in immunocompromised or otherwise debilitated human hosts. In recent years, a variety of species within the genus *Mycobacterium*, including several of the "atypical", rapidly growing strains that appear to be common on the RO membrane surfaces, have been implicated as disease agents in man [17, 33]. Moreover, there is increasing epidemiological evidence that many of these organisms may be transmitted by a waterborne route involving potable water distribution systems [17,25,26]. Disinfection of the feedwater and permeate streams of RO systems used in potable water applications would appear to be of special importance given the concerns outlined above. In addition, minimal sanitary precautions should probably be exercised by operations or maintenance personnel who might come into frequent contact with biofouled RO membrane surfaces, particularly if they are used in wastewater treatment applications.

Several key factors may favor the early exclusive attachment and growth of mycobacteria on the cellulose acetate RO membrane surfaces. For example, it is known that many mycobacteria are considerably more resistant than most other microbial types to the microbicidal effects of chlorine and other chemical disinfectants [17,28]. Thus, the mycobacteria may enjoy a strong selective advantage compared to other, more chlorine-sensitive microbes which adhere to the RO membrane surface. In addition to being more chlorine tolerant, the mycobacteria possess an extremely hydrophobic cell exterior due to high molecular weight peptidoglycolipids and waxy esters which comprise their cell walls [24,52]. Since the RO membrane surface also possesses a hydrophobic character, hydrophobic interactions may promote the attraction of these bacteria.

The mycobacteria, which comprise the nascent biofilm, are believed to biochemically modify the new, unfouled RO membrane surface via the *de novo* synthesis and extrusion of extracellular biopolymers (i.e., glycocalyx) used for the purpose of adhesion stabilization and nutrient acquisition [16,18-20,57,92]. Such progressive biochemical surface modification may, over time, render the RO membrane less recognizable to the mycobacteria and more susceptible to colonization by other less chlorine-resistant microbial forms. These "secondary colonizers" may, in turn, further modify the already biofouled membrane surface through their own metabolic activities, as well as by the production and deposition of additional glycocalyx materials, thereby perpetuating further successional changes in population structure.

Biofilm Structure

There have been relatively few detailed investigations of the ultrastructural characteristics of RO biofilms [2,69-71], which is unfortunate, since it is the fine-structural attributes of the biofilm that are perhaps most important in influencing the flux and ion permselectivity properties exhibited by the RO membrane. When examined by scanning electron microscopy (SEM), the biofilm which develops on the feedwater surface of an RO membrane can be seen to consist of predominantly rod-shaped bacteria embedded within an extensive glycocalyx matrix (Figure 10). The production of extracellular slime, consisting primarily of acidic heteropolysaccharides and glycoproteins [18-20,92], by the adherent microorganisms is frequently so extensive that the individual cells comprising the biofilm are occluded from view. It is the production of such glycocalyx material, and its extreme propensity for hydration, that imparts the slippery, gelatinous texture to biologically fouled RO membrane surfaces.

The submicroscopic packing arrangement and hydration capacity of the individual biopolymer strands produced by the adherent fouling bacteria represents two of the key factors which regulate the transport of water molecules across the RO membrane. Increased biopolymer packing densities and moisture affinity would presumably correlate with reduced membrane flux. Besides playing an important, if not essential, role in promoting and maintaining adhesion of the fouling bacteria within the biofilm [16,18-20], the bacterial glycocalyx is also known to be involved in the adsorption and concentration of soluble organic and inorganic nutrients near the cell surface [18,48,57]. Thus, the bacteria which constitute the biofilm are frequently exposed to elevated nutrient levels which, in turn, tends to enhance their metabolic activities and growth rates compared to their planktonic counter-

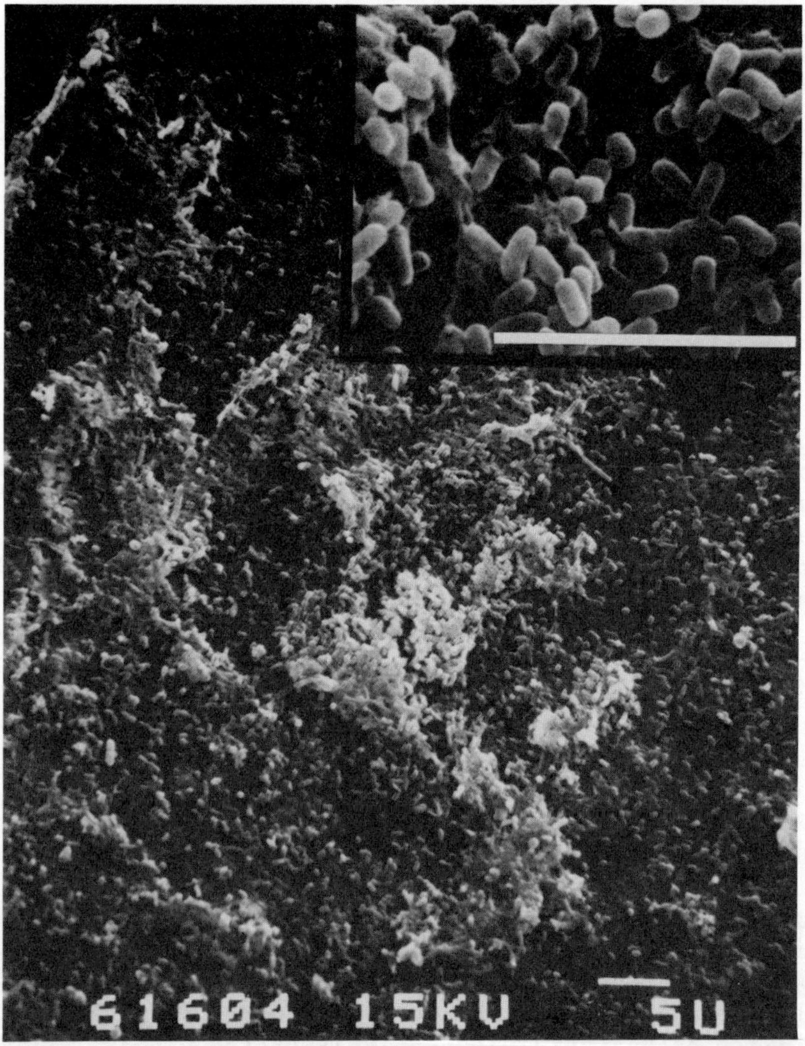

FIGURE 10. SEM photo of a mature biofilm on the feedwater surface of an RO membrane. Note fibrillar glycocalyx associated with adherent cells (inset). Bars = 5.0 μm.

parts [57].

The bacterial glycocalyx associated with RO biofilms is also important from the standpoint of RO membrane cleaning and system sanitization. It has been demonstrated experimentally that biopolymer synthesis and extrusion greatly increases the resistance of attached microorganisms to chlorine and other chemical disinfectants [18-20,84,85]. Indeed, this effect is so pronounced that once a surface biofilm is established, the resident bacteria become resistant to even a free chlorine residual of several parts per million [46,72,84]. The true fibrillar nature of the bacterial glycocalyx and the internal fine structure of the RO biofilm can best be viewed by transmission electron microscopy (TEM) of thin-sectioned specimens. As indicated in the TEM photomicrograph depicted in Figure 11, the mature RO biofilm possesses a distinct laminar construction, with each layer being composed predominantly of partially or completely autolysed bacteria devoid of cytoplasmic contents. Interspersed at random throughout the lytic cell debris were morphologically intact bacteria, many of which were firmly affixed within the biofilm matrix by means of a network of extracellular fibrils radiating outward from the cell surface. The compressed appearance of the biofilm layers is presumably the result of the hydraulic pressure differential applied during membrane operation. When the biofilm was viewed in an edgewise orientation, the laminar structure was also evident by SEM (Figure 12).

Bacterial colonization of the polyester, straight-woven Texlon fibers on the permeate surfaces of biofouled RO membranes has also been observed [69,71]. As is shown in Figure 13, such colonization was typically quite sparse, with large regions of the Texlon fibers remaining entirely barren. Like the fouling microbes attached to the feedwater surface of the RO membrane, bacteria on the permeate surface also appeared to be attached by means of extracellular

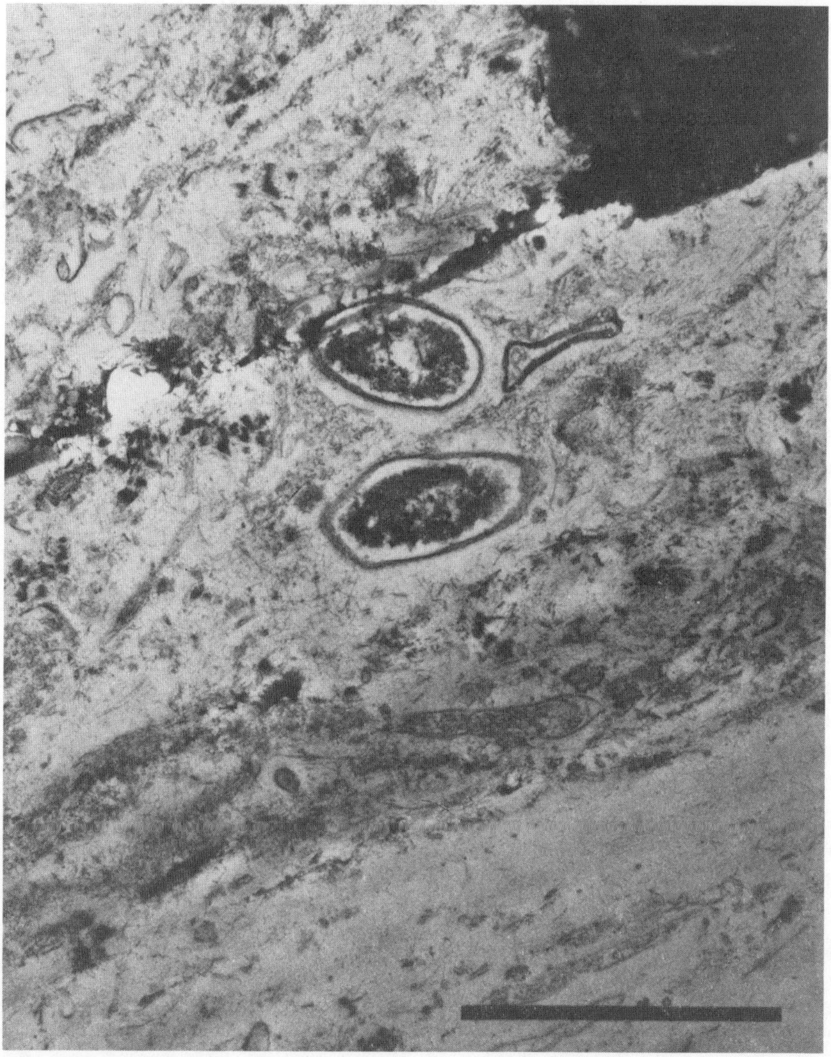

FIGURE 11. Thin-sectioned preparation of mature RO biofilm. Note autolysed and intact bacteria, as well as laminar structure of biofilm. Bar = 1.0 μm.

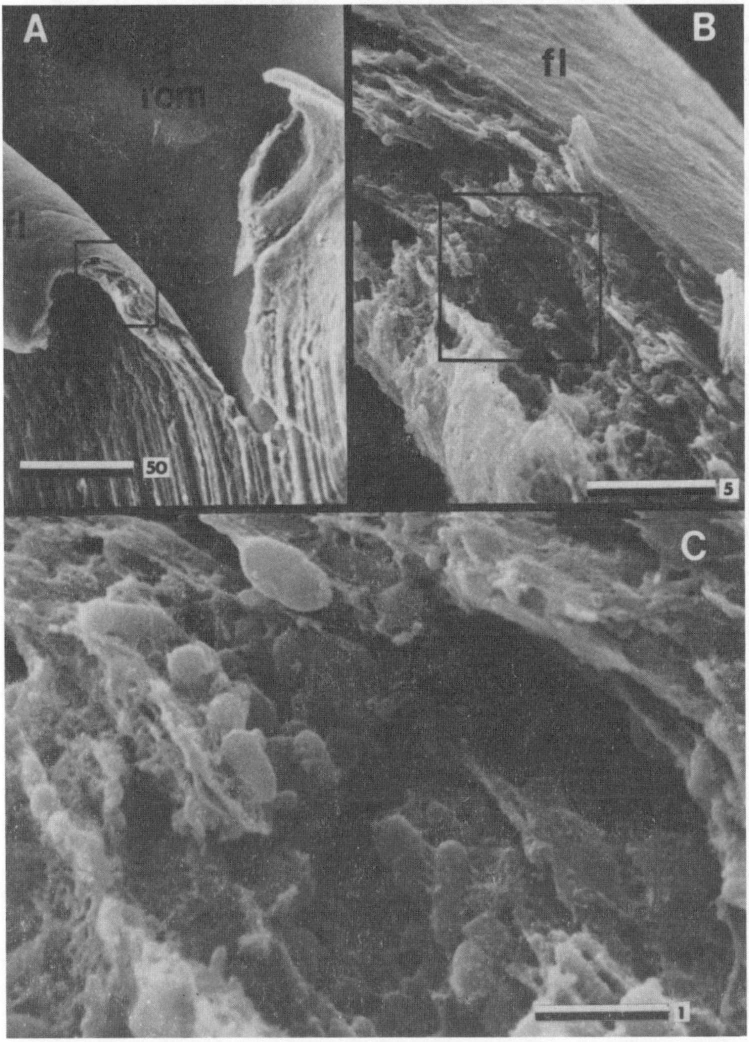

FIGURE 12. SEM photomicrograph of biofilm in edge view. Note layered appearence of biofilm and extra-cellular polymers used for adhesion and nutrient aquisition. Bar sizes indicated in μm. From Ridgway _et_ _al_. [69].

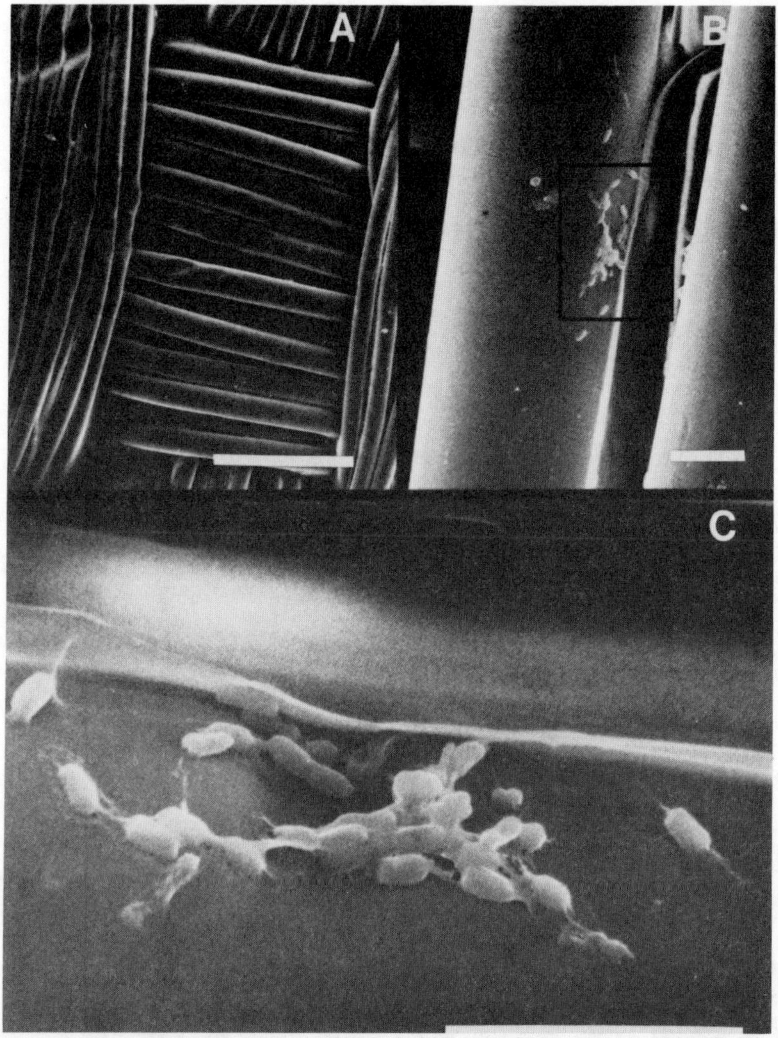

FIGURE 13. SEM photomicrographs depicting microbial colonization of polyester Texlon support fibers on permeate surface of RO membrane. Such colonization can result in contamination of downstream processes. Note polymeric glycocalyx material in lower panel. Bar sizes = 50 μm (panel A); 5.0 μm (panels B and C).

fibrillar material. Microbial colonization of the Texlon fibers, though sparse, is nevertheless significant since it could greatly increase the potential for microbial and/or particulate contamination of critical downstream processes. Moreover, the bacteria which colonize the Texlon fiber surfaces are already adapted to oligotrophic conditions, a factor which may enhance their survival and regrowth in permeate distribution lines or product-water storage tanks.

The origin of bacteria on the permeate surfaces of RO membranes is not known, although it is possible that they could have been introduced from the feedwater surface via microscopic holes or fissures in the membrane fabric. It is also possible that such microbes could have contaminated the Texlon fiber surfaces during the membrane manufacturing process, which is typically not performed under aseptic conditions. Prolonged storage or shipment of the assembled membrane modules in a moist or wetted state would tend to favor the growth and multiplication of microbial contaminants introduced at some earlier stage of membrane manufacture.

Effects of Biofouling on Membrane Performance

There are three principle effects of RO membrane biofouling, which include (1) a gradual decline in the permeability of the membrane to water (i.e., a membrane flux decline), (2) a gradual increase in membrane Δp, and (3) an increase in the permeability of the membrane to dissolved solids (i.e., a loss of mineral rejection capacity). Collectively, these factors translate into a significant reduction in the operating efficiency of the RO process, greatly increased operating and maintenance costs, and a shortened membrane life time.

The structure, composition, and hydration potential of the attached biofilm ultimately determines the rate and extent of membrane flux decline. Typically, a flux decline due to biofouling assumes biphasic kinetics: an early, rapid decline phase lasting several months to a year or more (depending on the specific operating conditions and feedwater quality), followed by a much slower decline phase lasting for the duration of operation [1,3,4,56]. The early flux loss usually follows linear or algebraic kinetics, whereas the later phase typically follows a linear decline (Figure 14). When biofouling occurs, a 30-50% decline in flux is not uncommon over the life of the membrane, which is generally two to three years.

In addition to a loss in net water permeation, fluid frictional resistance over the membrane surface is increased when biofouling occurs. The relationship between fluid frictional resistance and surface biofilm development has been studied intensively in a variety of model flow systems [8,13-15,50]. In an RO module, increased fluid drag resulting from membrane biofouling causes an increase in system Δp (Figure 14). When the Δp surpasses the maximum value that is recommended by the membrane manufacturer, there is a risk that the module will be permanently damaged, usually by a physical distortion or compression of the module. Often such distortion occurs as a result of "telescoping" of the membrane leaves within the element. Telescoping causes adjacent membrane leaves to shift position relative to one another, thereby increasing the likelihood of abrasion and rupture of the very thin, semipermeable film. Perhaps the most serious effect of biofouling in RO systems is a gradual decline in the mineral rejection capacity of the membranes (Figure 14). At least two independent mechanisms appear to be instrumental in the loss of mineral rejection as a result of membrane biofouling. First, certain fouling microorganisms may be capable of producing and excreting enzymes or other metabolites which can enhance the rate of hydrolysis

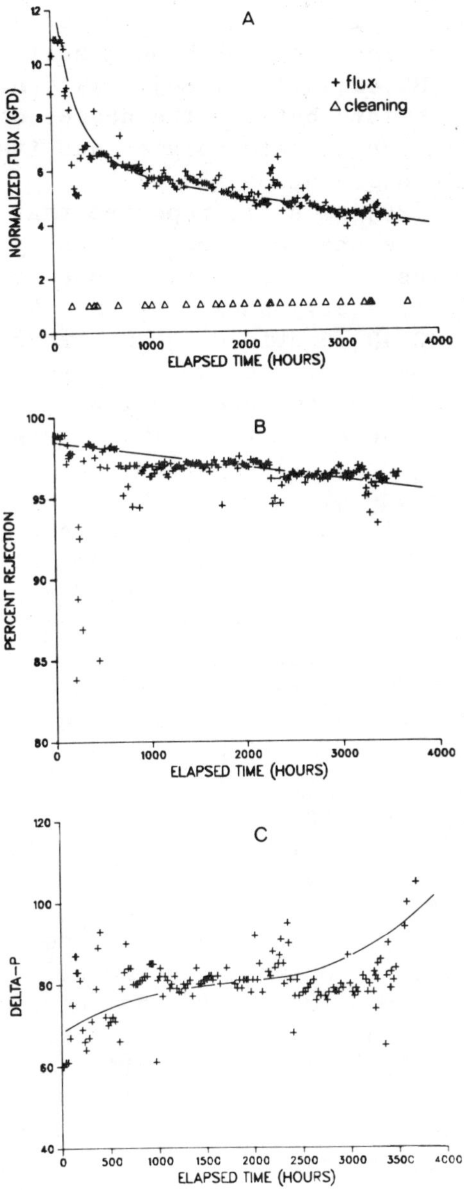

FIGURE 14. Effects of biofouling on performance of a polyamide RO membrane operated on pretreated municipal wastewater. Note biphasic flux decline (A), gradual loss of mineral rejection (B), and increase in membrane Δp (C). Note progressive flux decline in spite of regular membrane cleaning (Triangles in A).

of the RO membrane polymer. This is particularly true of cellulosic membrane polymers (e.g., cellulose acetate) for which numerous instances of biodegradation have been reported [11,37,42,43,45,54,86]. Evidently, a positive correlation exists between the degree of substitution of a cellulose acetate polymer and its potential for enzymatic degradation [39,67]. Accordingly, Ho _et al_. [39] recently reported that most fouling bacteria were unable to enzymatically hydrolyze modern cellulose acetate polymers due to the relatively high degree of acetyl substitution. Nevertheless, Kutz _et al_. [42,43] provided recent experimental evidence that a _Seliberia_ sp. was able to degrade cellulose acetate RO membranes following prolonged incubation in a mineral salts medium. In addition, Motomura and Taniguchi [54] reported the degradation of cellulose acetate RO membranes in areas which appeared to be localized directly beneath colonies of adherent fouling bacteria. However, no data were presented in either study which suggested that such degradation was the result of enzymatic processes. Thus, it is possible that the observed membrane deterioration was due to the localized accumulation of acidic or alkaline metabolites (e.g., organic amines) in proximity to the heavily biofouled regions of the membrane surface.

A second mechanism by which fouling microbes may adversely impact mineral rejection is by enhancing the concentration of dissolved ions at the RO membrane surface. As water preferentially passes through the RO membrane, solutes transported in the feedwater tend to become concentrated in a non-turbulent zone (i.e., the viscous boundary layer) proximal to the membrane surface. Such a "concentration polarization" effect occurs despite the turbulence-promoting activity of the plastic feed channel spacer (Vexar) which separates adjacent membrane leaves within the RO module. The formation of an attached biofilm markedly amplifies the effect of concentration polarization by extending the non-turbulent zone beyond that which is characteristic of a clean (i.e., unfouled) membrane surface. In the absence of adequate mixing at the membrane-solution interface, both organic and inorganic solutes can become highly concentrated within the biofilm matrix. Because solute transport through the RO membrane is proportional to the concentration gradient across the membrane [41,89], more salts tend to pass through the membrane into the permeate stream as the biofilm develops in thickness; hence, the apparent loss of mineral rejection with time. Presumably, this effect would be largely reduced or eliminated after membrane cleaning.

Pretreatment Methods

The two strategies that are most commonly employed to control biological fouling in RO systems include improvements in feedwater pretreatment and chemical cleaning of the membrane surfaces. Pretreatment measures typically include continuous or intermittent disinfection of the feedwater stream, and/or filtration to remove microorganisms and other colloidal particles which may promote membrane biofouling. The most effective and economical methods of chemical disinfection include chlorination (or chloramination), ozonation, and iodination of the RO feedwater. Unfortunately, many RO membrane polymers are irreversibly damaged (via halogenation and depolymerization reactions) by microbiocidally-active concentrations of such oxidizing disinfectants (Table 2). Various cellulose acetate and sulfonated polysulfone type membranes appear to be most resistant to the depolymerizing effects of relatively low concentrations of chemical oxidants [34,35,41,89]. However, polyamide and many thin-film composite type membranes are remarkably sensitive to these agents (with possible exceptions of iodine [63]), being unable to tol-

TABLE 2. Compatibility of cleaning compounds with different RO membranes.

	Compatibility with:		
Component or Condition	CA Membrane	PA Membrane	TFC Membrane
pH Range	2 - 8	2 - 12	2 -12
Maximum Temperature (F)	120	120	120
Detergents/Surfactants			
anionic (sodium dodecyl sulfate)	Y[1]	Y/N[2]	Y/N
cationic (cetyltrimethylammonium bromide; alkylpyridinium chloride)	Y	N[3]	N
ampholytic (dodecyl-N-betaine)	Y	Y/N	Y/N
nonionic (Triton X and Triton N series)	Y	N	N
Disinfectants/Biocides			
chlorine (free; less than 1 mg/l)	Y	N	N
chlorine (combined; less than 5 mg/l)	Y	N	N
chlorine dioxide	Y	Y/N	Y/N
ozone	N	N	N
hydrogen peroxide	Y/N	N	N
iodine	Y	Y/N	Y/N
formaldehyde/glutaraldehyde	Y	Y	Y
copper sulfate	Y	Y	Y
sodium bisulfite	Y	Y	Y
quaternary amines	Y	N	N
Chelating Agents			
EDTA/EGTA[4]	Y	Y	Y
citrate	Y	Y	Y
Chaotropic Agents			
urea	Y	Y/N	Y/N
guanidinium hydrochloride	Y/N	Y/N	Y/N
Enzymes			
proteases, lipases, esterases,etc.	Y	Y/N	Y/N

[1] Y = usually compatible

[2] Y/N = compatible in some cases, or compatibility unknown

[3] N = usually not compatible

[4] EDTA = ethylenediaminetetraacetic acid; EGTA = ethyleneglycol-bis(B-aminoethylether)N,N'-tetraacetic acid

erate even trace quantities of oxidizing disinfec tants in the feedwater supplying the RO system [34, 35]. In certain situations where the effects of chemical oxidants on downstream processes cannot be tolerated (e.g., in hemodialysis systems), it may be undesirable to use oxidizing disinfectants on a continuous basis, since they may readily penetrate the semipermeable membrane barrier and enter the permeate stream. In such instances, it is sometimes advisable to dechlorinate the feedwater immediately prior to its entry into the RO modules. Frequently employed methods of dechlorination include passage of the feedwater through granular activated carbon (GAC) adsorption columns, or stoichiometric addition of sodium thiosulfate, or some other appropriate chemical reducing agent (e.g., bisulfite), to the feed stream. The use of GAC columns prior to RO demineralization is sometimes less than an ideal situation, however, since many kinds of microorganisms, including a variety of opportunistic pathogens, adhere to and extensively colonize the carbon particles themselves [9,32,46]. Eventually, these bacteria begin to slough off the carbon in large numbers and enter the RO feedwater where they are transported inadvertantly onto the RO membrane surfaces. Indeed, the viable bacterial concentration in the effluent stream from many GAC systems is often ten to one hundred-fold higher than that of the influent water [32,69].

Despite whatever other disadvantages may be associated with the use of oxidizing disinfectants and the precautions that must be exercised in their application, they continue to represent the most reliable and economical method for microbial disinfection and biofouling control in RO systems. The principle advantages of chlorine and similar chemical disinfectants is that they are (1) comparatively inexpensive, (2) extremely biocidal and effective against a wide range of microbial types, (3) provide a continuously available residual on the RO membrane surface where biofouling occurs, (4) can be conveniently inactivated if necessary by the addition of a suitable reducing agent or by passage through a GAC column, and (5) can be readily washed out of a system following disinfection.

Perhaps the most frequently employed alternate method of disinfection for the control of biofouling in RO systems is ultraviolet (UV) irradiation of the feedwater stream. This technique has been successfully employed in numerous installations and has the advantage of being a relatively self-contained, low maintenance type of system that will not damage the RO membrane. However, many types of bacteria that are commonly found in domestic water supplies are known to be extremely resistant to UV irradiation. Furthermore, unlike treatment by chemical disinfectants, UV irradiation of the feedwater fails to provide a continuously available disinfectant residual on the RO membrane surface, where it is most needed. Presumably, bacteria which survive UV irradiation would still be capable of rapid growth and proliferation on the RO membrane surface. In contrast, a viable microorganism that adheres to an RO membrane supplied with a chlorinated feed water would be continuously exposed to the disinfectant, thereby greatly increasing the likelihood of its eventual inactivation.

Prefiltration of the RO feedwater may also be utilized to physically remove bacteria and other suspended particulate matter from the feedwater before they come into contact with the RO membrane surface. Unfortunately, most standard RO prefilters possess nominal or absolute pore sizes ranging from 5 to 25 μm in diameter, which is, of course, far too large to efficiently retain bacteria and prevent their entrance into the RO module. Consequently, only submicron filters or ultrafilters can be successfully utilized for this purpose. However, the economics of ultrafiltration as a pretreament process are often questionable, especially in larger RO systems where flux losses assume more of an economic significance. Consequently, their use depends largely on the colloidal and microbiological composition and fouling characteristics of the RO feedwater, with higher bacterial concentrations and turbidities resulting in accelerated plugging of the filters. If the ultrafilters plug too rapidly, the frequency and duration of their cleaning may be so great as to offset any economic advantage gained by the reduction in biofouling of the RO membranes.

Membrane Cleaning

One of the most effective methods for the conrol of microbial fouling in RO systems is regular chemical cleaning of the membrane surfaces. Virtually all of the commerically available cleaning formulations have been developed by empirical, trial-and-error evaluation. Rigorous testing and evaluation of commercial membrane cleaning solutions under controlled laboratory conditions have not been performed. Therefore, it is not possible to provide an accurate comparative analysis of the efficiencies of each cleaning formulation in terms of biofilm removal and restoration of RO membrane performance.

Clearly, it is the bacterial adhesion process that ultimately determines the tenacity and structural integrity of the RO biofilm. Thus, it is crucial to have a concise understanding of the subcellular and molecular mechanisms which mediate bacterial adherence to RO membranes before truly effective strategies for biofilm removal, or inhibition of biofilm formation, can be successfully developed. Chemical agents which interfere with or otherwise disrupt the hydrophobic and ionic interactions between specific bacterial cell surface components (e.g., glycocalyx, fimbria, etc.) and the RO membrane polymer will be most likely to effect good biofilm removal.

Despite current limitations in our understanding of the bacterial adhesion process, it is instructive to examine some possible strategies that might be employed in the formulation of a cleaning solution for biofilm removal. Some general strategies for biofilm removal are diagrammatically illustrated in Figure 15.

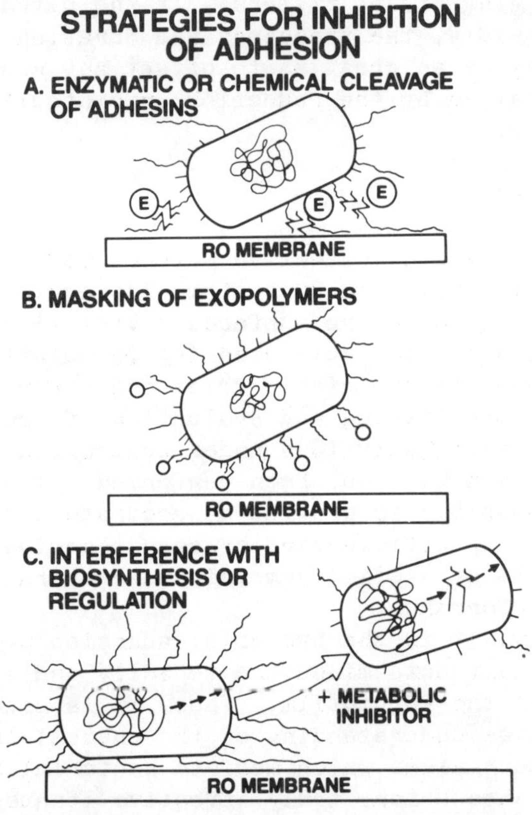

FIGURE 15. Diagram illustrating some general strategies for biofilm removal, including enzymatic cleavage of biopolymers used for adhesion (A), detergent masking of hydrophobic cell surface biopolymers involved in adhesion (B), and inhibition of biopolymer synthesis by metabolic inhibitors (C).

Included among these strategies are (1) the utilization of enzymes or other chemical agents which can directly attack and cleave the extracellular biopolymers which help mediate bacterial adhesion to the RO membrane surfaces, (2) the application of specific antimetabolites or biocidal agents which can interfere with biopolymer synthesis and/or extrusion, and (3) the addition of surface-active agents which can mask hydrophobic groups on the cell surface, solubilize biological membranes [23,38,53,80,81], and aid in the penetration of chemical biocides into the biofilm matrix. Because hydrophobic interactions appear to be involved in the initial attachment of certain fouling bacteria to RO membrane surfaces [74, 75], the addition of surfactants to many cleaning formulations seems logical, and indeed has generally proved to be effective at removing some biofilms (Table 3).

The exploration and testing of new detergents for biofilm removal is an area of research that still holds a great deal of promise. For example, very recent studies have indicated that many anionic detergents that were once believed to be relatively non-biocidal, become strongly antibacterial when utilized synergistically in combination with lanthanum salts, or the salts of other rare-earth metal cations [7]. However, it should be cautioned that the interaction of surfactants with biological membranes and other cellular structures is a highly complex phenomenon which is rarely predictable [38]. As was pointed out in an earlier section of this chapter, some surfactants (such as certain of the cationic quaternary ammonium detergents) may actually promote bacterial adhesion to the RO membrane surface depending upon their molecular structure and charge characteristics. It should also be noted that detergents possessing strongly anionic, cationic, or amphipathic side groups may be incompatible with some RO membrane types, especially certain polyamide and

TABLE 3. Roles of common RO cleaner constituents

Component	Purpose/Function	Typical Examples	Effective Concentration Range
Detergent/ Surfactant	Disruption and/or solubilization of biological membranes or membrane components; assists penetration of chemical biocides into the biofilm	Triton X or Triton N series sodium dodecyl benzene sulfonate alkyltrimethylammonium bromide trisodium phosphate; sodium tripolyphosphate	0.01 - 0.1%
Chelating Agents	Sequesters divalent cations involved in the stabilization of the bacterial cell wall	EDTA, EGTA citrate	0.001 - 0.1% 0.1 - 1.0%
Chaotropic/ Denaturing Agents	Disrupts the secondary and tertiary structure of cell proteins which may be involved in bacterial adhesion and biofilm stabilization	urea guanidinium hydrochloride	40 - 60% 10 - 30%
Biocidal Agents	metabolically inactivates fouling bacteria	chlorine and other oxidizing disinfectants formaldehyde quaternary amines	0.5-5.0 ppm 0.5 - 5.0% 0.01 - 0.1%
Enzymes	Catalyzes the cleavage of specific chemical bonds in extracellular biopolymers which are involved in maintaining biofilm integrity	esterases lipases proteases	10-500 ppm " " " " " "

thin-film composite membranes, causing a rapid and sometimes irreversible flux decrease following treatment (Table 2). Nonionic, polyoxyethylene type detergents, such as the Triton X, Triton N, or Nonidet P40 series surfactants, may exhibit the widest degree of compatibility with different RO membrane polymers. However, even the nonionic surfactants must be used with extreme care when formulating membrane cleaning solutions, since low-purity preparations of some of these compounds (e.g., Triton X-100) have been shown to contain significant concentrations of peroxides, which have the potential to harm polyamide and thin-film composite type membranes [5,12].

Whittaker _et_ _al_. [91] employed SEM to qualitatively screen the effectiveness of various chemical and enzymatic agents for their ability to remove biofilm from cellulose acetate RO membranes. Among the most effective agents for biofilm removal were certain denaturant/detergent combinations (e.g., urea/sodium dodecyl sulfate). Subsequent experimentation demonstrated that six molar urea alone, a strong protein denaturing compound, tended to partially solubilize the biofilm [Ridgway, unpublished data]. Such observations suggest the involvement of bacterial surface proteins in maintaining the structure and integrity of the RO biofilm. Cell surface polypeptides have similarly been implicated in the initial adhesion of bacteria to other types of solid substrates, including glass and metal surfaces [31,48,49].

Other factors which are frequently exploited to enhance biofilm removal and restore the membrane flux to prefouling levels include (1) the use of higher cleaning temperatures, which tends to improve the solubilizing activity of most surfactant molecules; (2) the use of divalent cation chelating agents, such as ethylenediaminetetraacetic acid (EDTA), for the sequestration of calcium and magnesium ions, which are known to be involved in the stabilization of cell wall architecture in many bacteria [23,80]; (3) the use of quaternary amines, copper sulfate, formaldehyde, or other microbicidal agents to suppress microbial growth on the membrane surfaces; and (4) the use of strongly alkaline pH values during cleaning (of polyamide or polysulfone membranes only), which acts to further denature cellular proteins involved in bacterial adhesion.

The restorative effects of membrane cleaning are usually transient at best. Although most of the original membrane flux typically can be recovered following cleaning, the membrane surfaces often undergo biofouling even more rapidly subsequent to the cleaning. The underlying reasons for accelerated biofouling and flux decline after cleaning are not altogether clear, although incomplete biofilm removal and more rapid *de novo* bacterial adhesion to an already biochemically "conditioned" membrane surface may promote re-fouling. In order to minimize the effect of refouling, it is probably best to clean the RO membranes at frequent and regular intervals (e.g., a minimum of every thirty days), before an extensive biofilm has had an opportunity to develop. Recent experimental evidence suggests that RO membrane cleaning solutions become noticeably less effective as a function of biofilm age [2,91].

SUMMARY AND CONCLUDING REMARKS

Reverse osmosis membranes become biologically fouled when bacteria and other microorganisms (e.g., fungi), which are transported in the feedwater, attach and proliferate at the membrane-liquid interface. Both hydrophobic and ionic interactions appear to be important in attracting and maintaining adhesion of the fouling bacteria to the RO membrane surfaces. The membrane biofilm develops rapidly, especially when the RO system is supplied by feedwater containing high bacterial numbers; and, in certain instances, may involve the extensive colonization of the RO membrane surface by opportunistic pathogens of public health concern, such as the acid-fast mycobacteria. The development of a microbial biofilm adversely affects the performance characteristics of the RO membrane, causing a gradual flux decline, a loss of mineral rejection, and an increase in the differential operating pressure of the system. The densely packed, laminar structure and microbial composition of the biofilm, as well as the metabolic activities of the individual resident bacteria, all appear to contribute to the observed membrane performance decline, which becomes progressively irreversible with time.

Because bacterial adhesion is a necessary prerequisite to the development and maintenance of the nascent biofilm, it is imperative that a clear understanding of the subcellular and molecular mechanisms that govern the adhesion process be acquired before rational strategies for biofouling control can be successfully developed and evaluated. Unfortunately, fundamental knowledge of the physiology and biochemistry of the bacterial adhesion process as it applies to RO biofouling remains sketchy at best. Currently employed remedial efforts for the control of RO biofouling, which involve feedwater pretreatment (e.g., filtration and disinfection) and membrane cleaning, are logical approaches. Yet they are often inadequate or too costly and provide only transient or partial recovery of membrane performance. Additional research and development are badly needed to explore the possibility of creating novel inhibitors of the bacterial adhesion process which could be utilized to help prevent biofouling in RO systems.

ACKNOWLEDGMENTS

The author wishes to thank Ms. Jana Safarik for her expert technical assistance. Thanks are also due to Mr. Kim Rutledge for the excellent technical illustrations. The cooperation of Pfizer Corporation, Groton, Connecticut, and Buckman Laboratories, Inc., Memphis, Tennessee are acknowledged.

Whereas a portion of the research described in this chapter was sponsored by the Orange County Water District, Pfizer Corporation, and Buckman Laboratories, Inc., the findings and conclusions do not necessarily reflect the individual or collective views of these organizations, and no official endorsement should be inferred.

References

1. Argo, D.G. 1980. Evaluation of membrane processes and their role in wastewater reclamation, Vol. II, U. S. Department of the Interior, Office of Water Research and Technology, Washington, D.C.

2. Argo, D.G. and H.F. Ridgway. 1982. Biological fouling of reverse osmosis membranes at water factory 21. In Proceedings of the Water Supply Improvement Association, Vol. 1, Honolulu, Hawaii.

3. Argo, D.G. and M.G. Rigby. 1981. Evaluation of membrane processes and their role in waste-water reclamation. Vol. III, U.S. Department of the Interior, Office of Water Research and Technology, Washington, D.C.

4. Argo, D.G., M.G. Rigby and H.F. Ridgway. 1984. Improvements in wastewater demineralization with advancing membrane technology. In Proceedings of the Water Supply Improvement Association, Vol. 1, Orlando, Florida, 1984.

5. Ashani, Y. and G.N. Catravas. 1980. Highly reactive impurities in Triton X-100 and Brij-35: partial characterization and removal. Anal. Biochem. 109:55-62.

6. Belas, M.R. and R.R. Colwell. 1982. Adsorption kinetics of laterally and polarly flagellated _Vibrio_. J. Bacteriol. 151:1568-1580.

7. Berg, R.W. and R.E. Zimmerer. 1987. Effect of rare earth cations on bactericidal activity of anionic surfactants. J. Ind. Microbiol. 1: 377-381.

8. Bryers, J. and W.G. Characklis. 1981. Early biofilm formation in a turbulent flow system: overall kinetics. Water Res. 15:483-491.

9. Burlingame, G.A., I.H. Suffet and W.O. Pipes. 1986. Predominant bacterial genera in granular activated carbon water treatment systems. Can. J. Microbiol. 32:226-230.

10. Cadotte, J.E. and L.T. Rozelle. 1972. _In-situ_ formed condensation polymers for reverse osmosis membranes, Contract No. 14-30-2883, U. S. Department of the Interior, Office of Saline Water Research and Development report, North Star Research and Development Institute.

11. Cantor, P.A. and B.J. Mechalas. 1969. Biological degradation of cellulose acetate reverse-osmosis membranes. J. Polymer Sci. 28:225-241.

12. Chang, H.W. and E. Bock. 1980. Pitfalls in the use of commercial nonionic detergents for the solubilization of integral membrane proteins: sulfhydryl oxidizing contaminants and

their elimination. Anal. Biochem. 104:112-117.

13. Characklis, W.G. 1973. Attached microbial growths-I. attachment and growth. Water Res. 7:1113-1127.

14. Characklis, W.G. 1973. Attached microbial growths-II. frictional resistance due to microbial slimes. Water Res. 7:1249-1258.

15. Characklis, W.G. and K.E. Cooksey. 1983. Biofilms and microbial fouling. Adv. Appl. Microbiol. 29:93-136.

16. Cheng, K.J., R.T. Irvin and J.W. Costerton. 1981. Autochthonous and pathogenic colonization of animal tissues by bacteria. Can. J. Microbiol. 27:461-490.

17. Collins, C.H., J.M. Grange and M.D. Yates. 1984. Mycobacteria in water. J. Appl. Bacteriol. 57:193-211.

18. Costerton, J.W. 1984. Mechanisms of microbial adhesion to surfaces. p. 115-123. In M.J. Klug and C.A. Reddy (ed.), Current perspectives in microbial eclolgy,American Society for Microbiology, Washington, D.C.

19. Costerton, J.W., R.T. Irvin and K.J. Cheng. 1981. The bacterial glycocalyx in nature and disease. Ann. Rev. Microbiol. 35:299-324.

20. Costerton, J.W., T.J. Marrie and K.J. Cheng. 1985. Phenomena of bacterial adhesion. p. 3-43. In D.C. Savage and M. Fletcher (ed.), Bacterial adhesion, mechanisms and physiological significance, Plenum Press, New York.

21. Crossley, I.A. 1983. Desalination by reverse osmosis. p. 205-248. In A. Porteous (ed.), Desalination technology, developments and practice, Applied Science Publisher, London.

22. Dahlback, B., M. Hermansson, S. Kjelleberg and B. Norkrans. 1981. The hydrophobicity of bacteria--an important factor in their initial adhesion at the air-water interface. Arch. Microbiol. 128:267-270.

23. DePamphilis, M. L. 1971. Dissociation and reassembly of *Escherichia coli* outer membrane and of lipopolysaccharide, and their reassembly onto flagellar basal bodies. J. Bacteriol. 105: 1184-1199.

24. Draper, P. 1982. The anatomy of mycobacteria. p. 9-48. In C. Ratledge and J. Stanford (ed.), The biology of the mycobacteria, Vol. 1, physiology, identification and classification, Academic Press, London.

25. DuMoulin, G.C., I.H. Sherman, D.C. Hoaglin, and K.D. Stottmeir. 1985. *Mycobacterium avium* complex, an emerging pathogen in Massachusetts. J. Clin. Microbiol. 22:9-12.

26. DuMoulin, G.C. and K.D. Stottmeir. 1986. Waterborne mycobacteria: an increasing threat to health. ASM News 52:525-529.

27. Eisenberg, T.N. and E.J. Middlebrooks. 1986. Reverse Osmosis of Drinking Water, Butterworth Publishers, Boston.

28. Englebrecht, R.S., C.N. Haas, J.A. Shular, D. L. Dunn, D. Roy, A. Lalchandani, B.F. Severin

and S. Farooq. 1979. Acid-fast bacteria and yeasts as indicators of disinfection efficiency, EPA-IAG-D6-0432, U. S. Environmental Protection Agency, Cincinnati, Ohio.

29. Fattom, A. and M. Shilo. 1984. Hydrophobicity as an adhesion mechanism of benthic cyano-bacteria. Appl. Environ. Microbiol. 47:135-143.

30. Fletcher, M. and G.I. Loeb. 1979. Influence of substratum characteristics on the attachment of a marine pseudomonad to solid surfaces. Appl. Environ. Microbiol. 37:67-72.

31. Fletcher, M. and S. McEldowney. 1984. Microbial attachment to nonbiological surfaces. p. 115-123. _In_ M.J. Klug and C.A. Reddy (ed.), Current perspectives in microbial ecology, American Society for Microbiology, Washington, D.C.

32. Geldreich, E.E., R.H. Taylor, J.C. Blannon and D.J. Reasoner. 1985. Bacterial colonization of point-of-use water treatment devices. J. Am. Water Works Assoc. 77:72-80.

33. Good, R.C. 1985. Opportunistic pathogens in the genus _Mycobacterium_. Ann. Rev. Microbiol. 39:347-369.

34. Glater, J.; J.W. McCutchan, S.B. McCray and M.R. Zachariah. 1981. The effect of halogens on the performance and durability of reverse-osmosis membranes. p. 171-190. _In_ A.F. Turbak (ed.), Synthetic membranes: vol. I, desalination, American Chemical Society Symposium Series, no. 153, American Chemical Society, Washington, D.C.

35. Glater, J., M.R. Zachariah, S.B. McCray and J.

W. McCutchan. 1983. Reverse osmosis membrane sensitivity to ozone and halogen disinfectants. Desalination 48:1-16.

36. Gregor, H.P. and C.D. Gregor. 1978. Membrane technology. Sci. Am. 239:88-101.

37. Guy, D.G. 1980. Biodegradation of cellulosic membranes at the Roswell and Yuma desalination test facilities. p. 12-15. In Pure water, Vol. 2, International Desalination and Environmental Association, Teaneck, New Jersey.

38. Helenius, A. and K. Simons. 1975. Solubilization of membranes by detergents. Biochim. Biophys. Acta 415:29-79.

39. Ho, L.D.W., D.D. Martin and W.C. Lindemann. 1983. Inability of microorganisms to degrade cellulose acetate reverse osmosis membranes. Appl. Environ. Microbiol. 45:418-427.

40. Kaakien, J.W. and C.D. Moody. 1984. Characteristics of reverse osmosis membrane fouling at the Yuma test facility. In Proceedings of the symposium on reverse osmosis and ultrafiltration, American Chemical Society, Philadelphia.

41. Pesting, R.E. 1971. Synthetic polymeric membranes. McGraw Hill Book Company, New York.

42. Kutz, S.M., D.L. Bentley, N.A. Sinclair and L.M. Kelley. 1986. Morphological diversity of a bacterium resmbling Seliberia: an electron microscopic evaluation of nutritional effects. Abstracts of the Annual Meeting of the American Society for Microbiology, Washington, D.C., p. 189.

43. Kutz, S.M., N.A. Sinclair and D.L. Bentley. 1985. Characterization of Seliberia and its effects on reverse osmosis membranes. Abstracts of the American Society for Microbiology, Las Vegas, Nevada, p. 223.

44. Langmuir, I. 1918. The adsorption of gases on plane surfaces of glass, mica, and platinum. J. Am. Chem. Soc. 40:1361-1401.

45. Leahy, T.M. 1980. Recent experience with CTA hollow fiber membranes at the Roswell test facility. In Proceedings of the 8th annual conference of the Water Supply Improvement Association, Ipswich, Massachusetts.

46. LeChevallier, M.W., T.S. Hassenaver, A.K. Camper and G.A. McFeters. 1984. Disinfection of bacteria attached to granular activated carbon. Appl. Eniviron. Microbiol. 48:18-923.

47. Loeb, S. 1981. The Loeb-Sourirajan membrane: how it came about. p. 1-9. In Synthetic membranes, Vol. 1, desalination, American Chemical Society symposium series, No. 153, American Chemical Society, Washington, D. C.

48. Marshall, K.C. 1976. Interfaces in microbial ecology. Harvard Unversity Press, Cambridge, Massachusetts.

49. Marshall, K.C. 1985. Mechanisms of bacterial adhesion at solid-water interfaces. p. 133-161. In D.C. Savage and M. Fletcher (ed.), Bacterial adhesion, mechanisms and physiological significance, Plenum Press, New York.

50. McCoy, W.F., J.D. Bryers, J. Robbins and J.W. Costerton. 1981. Observations of fouling bio-

film formation. Can. J. Microbiol. 27:910-917.

51. McCutchan, J.W. and J.S. Johnson. 1970. Reverse osmosis at Coalinga, California. J. Am. Water Works Assoc. 62:346-353.

52. Minnikin, D.E. 1982. Lipids: complex lipids, their chemistry, biosynthesis and roles. p. 95-175. In C. Ratledge and J. Staford (ed.), The biology of the mycobacteria, Vol. 1, Physiology, identification, and classification, Academic Press, London.

53. Moriyon, I. and D.T. Berman. 1982. Effects of nonionic, ionic, and dipolar ionic detergents and EDTA on the Brucella cell envelope. J. Bacteriol. 152:822-828.

54. Motomura, H. and Y. Taniguchi. 1981. Durability study of cellulose acetate reverse osmosis membranes under adverse circumstances for desalting. p. 79-88. In A.F. Turbak (ed.), Synthetic membranes, Vol. 1: desalination. American Chemical Society symposium series no. 153, American Chemical Society, Washington, D.C.

55. Nir, S. 1983. Van der Waals interactions between surfaces of biological interest. Prog. Surf. Sci. 13:1-58.

56. Nusbaum, I. and D.G. Argo. 1984. Design, operation, and maintenance of a 5-mgd wastewater reclamation reverse osmosis plant. p. 377-436. In G. Belfort (ed.), Synthetic membrane processes, fundamentals and water applications, Academic Press, Orlando, Florida.

57. Paerl, H.W. 1985. Influence of attachment on microbial metabolism and growth in aquatic eco-

systems. p. 363-400. In D.C. Savage and M. Fletcher (ed.), Bacterial adhesion, mechanisms and physiological significance, Plenum Press, New York.

58. Parrett, T. 1984. Membranes: succeeding by separating. Technol. 3:16-29.

59. Paul, J.H. 1984. Effects of antimetabolites on the adhesion of an estuarine *Vibrio* sp. to polystyrene. Appl. Environ. Microbiol. 48:924-929.

60. Paul, J.H. and W.H. Jeffrey. 1985. Evidence for separate adhesion mechanisms for hydrophilic and hydrophobic surfaces in *Vibrio proteolytica*. Appl. Environ. Microbiol. 50:431-437.

61. Petersen, R.J. 1986. The expanding roster of commercial reverse osmosis membranes. Proceedings of the 1st biennial conference of the National Water Supply Improvement Association, Washington, D.C.

62. Petrocci, A.N. 1983. Surface-active agents: quaternary ammonium compounds. p. 309-329. In S. Block (ed.), Disinfection, sterilization, and preservation, Lea and Febiger, Philadelphia.

63. Pohland, H.W. and G.E. Bettinger. 1981. Successful operation of a perasep permeator reverse osmosis system on biologically active feed water. p.399-406. In A.F. Turbak (ed.), Synthetic membranes, Vol. 1, Desalination, American Chemical Society symposium series, No. 153, American Chemical Society, Washington, D.C.

64. Probstein, R.F., K.K. Chan, R. Cohen and I. Rubenstein. 1981. Model and preliminary experiments on membrane fouling in reverse osmosis. p.

131-145. In A.F. Turbak (ed.), Synthetic membranes, Vol. 1: desalination, American Chemical Society symp. series, No. 153, American Chemical Society, Washington, D.C.

65. Pusch, W. and A. Walch. 1982. Synthetic membranes, preparation, structure, and application. Angewandte Chemie 21:660-685.

66. Reed, R.H. and G. Belfort. 1982. Characterization of fouling potential for pressure-driven membrane processes: a new simulated flow cell. Water Sci. Technol. 14:499-522.

67. Reese, E.T. 1957. Biological degradation of cellulose derivatives. J. Indust. Eng. Chem. 49:89-93.

68. Reinhard, M., N.L. Goodman, P.L. McCarty and D.G. Argo. 1986. Removing trace organics by reverse osmosis using cellulose acetate and polyamide membranes. J. Am. Water Works Assoc. 78:163-174.

69. Ridgway, H.F., D.G. Argo and B.H. Olson. 1981. Factors influencing biofouling of reverse osmosis membranes at Water Factory 21: chemical, microbiological, and ultrastructural characterization of the fouling layer, Vol. III B., U. S. Department of the Interior, Office of Water Research and Technology, Washington, D.C.

70. Ridgway, H.F., C.A. Justice, C. Whittaker, D. G. Argo and B.H. Olson. 1984. Biofilm fouling of RO membranes--its nature and effect on treatment of water for reuse. J. Am. Water Works Assoc. 76:94-102.

71. Ridgway, H.F., A. Kelly, C. Justice and B.H.

Olson. 1983. Microbial fouling of reverse osmosis membranes used in advanced wastewater treatment technology: chemical, bacteriological, and ultrastructural analysis. Appl. Environ. Microbiol. 45:1066-1084.

72. Ridgway, H.F. and B.H. Olson. 1982. Chlorine resistance patterns of bacteria from two drinking water distribution systems. Appl. Environ. Microbiol. 44:72-987.

73. Ridgway, H.F., M.G. Rigby and D.G. Argo. 1984. Adhesion of a _Mycobacterium_ sp. to cellulose diacetate membranes used in reverse osmosis. Appl. Environ. Microbiol. 47:61-67.

74. Ridgway, H.F., M.G. Rigby and D.G. Argo. 1985. Bacterial adhesion and fouling of reverse osmosis membranes. J. Am. Water Works Assoc. 77:97-106.

75. Ridgway, H.F., M.G. Rigby and D.G. Argo. 1984. Biological fouling of reverse osmosis membranes: the mechanism of bacterial adhesion. _In_ Proceedings of the water reuse symposium III, The future of water reuse, San Diego, California, Vol. III, p. 1314-1350.

76. Ridgway, H.F., D.M. Rogers and D.G. Argo. 1986. Effect of surfactants on the adhesion of mycobacteria to reverse osmosis membranes. p. 133-164. _In_ Proceedings of the semiconductor pure water conference, San Francisco, California.

77. Riley, R.L. 1986. Thin-film composite reverse osmosis membranes: development needs and opportunities. _In_ Proceedings of the conference on technology and planning, Cambridge,

Massachusetts.

78. Safarik, J., D.M. Rodgers and H.F. Ridgway. 1987. Effects of biocidal agents on bacterial adhesion to reverse osmosis membranes. Abstracts of the annual meeting of the American Society for Microbiology, Atlanta, Georgia, p. 250.

79. Salit, I. E. 1983. Effect of subinhibitory concentrations of antimicrobials on meningococcal adherence. Can. J. Microbiol. 29:369-376.

80. Schnaitman, C.A. 1971. Effect of ethylenediaminetetraacetic acid, Triton X-100, and lysozyme on the morphology and chemical composition of isolated cell walls of _Escherchia coli_. J. Bacteriol. 108:553-563.

81. Schnaitman, C.A. 1971. Solubilization of the cytoplasmic membrane of _Escherichia coli_ by Triton X-100. J. Bacteriol. 108:545-552.

82. Scott, J. (ed.). 1980. Membrane and ultrafiltration technology, recent advances. Noyes Data Corporation, Park Ridge, New Jersey.

83. Scott, J. (ed.). 1981. Desalination of seawater by reverse osmosis. Noyes Data Corporation, Park Ridge, New Jersey.

84. Seyfried, P.L. and D.J. Fraser. 1980. Persistence of _Pseudomonas aeruginosa_ in chlorinated swimming pools. Can. J. Microbiol. 26:350-355.

85. Shabtai, Y. and D.L. Gutnick. 1985. Tolerance of _Acinetobacter calcoaceticus_ to the cationic surfactant cetyltrimethylammonium bromide: role of the bioemulsifier emulsan. Appl. Environ.

Microbiol. 49:192-197.

86. Sinclair, N.A. 1982. Microbial degradation of reverse osmosis desalting membranes. In Operation and maintenance of the Yuma desalting test facility, Vol. IV, U.S. Department of the Interior, Bureau of Reclamation, Washington D.C.

87. Smith, B.R. 1981. Fouling in whey reverse osmosis. p. 37-44. In A.F. Turbak (ed.), Synthetic membranes, Vol. II, hyper- and ultrafiltration uses, American Chemical Society symposium series, No. 154, American Chemical Society, Washington, D.C.

88. Torrey, S. (ed.). 1984. Membrane and ultrafiltration technology, developments since 1981, Noyes Data Corporation, Park Ridge, New Jersey.

89. U.S. Department of the Interior. 1979. Reverse osmosis technical manual, Office of Water Research and Technology, U.S.Department of the Interior, Washington, D.C.

90. Wechsler, R. 1976. Reverse osmosis on secondary effluent: the effect of recovery. Water Res. 11:379-385.

91. Whittaker, C., H.F. Ridgway and B.H. Olson. 1984. Evaluation of cleaning strategies for removal of biofilms from reverse osmosis membranes. Appl. Environ. Microbiol. 48:395-403.

92. Wicken, A.J. 1985. Bacterial cell walls and surfaces. p. 45-70. In D.C. Savage and M. Fletcher (ed.), Bacterial adhesion, mechanisms and physiological significance, Plenum Press, New York.

93. Winfield, B.A. 1979. The treatment of sewage effluents by reverse osmosis-pH based studies of the fouling layer and its removal. Water Res. 13:561-564.

94. Winfield, B.A. 1979. A study of the factors affecting the rate of fouling of reverse osmosis membranes treating secondary sewage effluents. Water Res. 13:565-570.

CHAPTER 7. BIOLOGICAL FOULING OF PURIFIED WATER SYSTEMS

Marc W. Mittelman, Water Micro Associates, San Diego, CA 92128

Introduction

Early man probably recognized the utility of purifying water when he noted that the water tasted better when the turbidity was allowed to settle out. Modern methods of water purification utilize a number of technologies including membrane filtration, distillation, reverse osmosis, ultrafiltration, deionization, electrodialysis reversal, ultraviolet light, ozonation, and carbon filtration. All of these methods--old and new--are designed to remove or otherwise inactivate some unwanted component from the feed water. Depending upon the intended use, various particulates, organics, ionic species, and microorganisms are considered as contaminants and are removed by one or more of these purification processes. The removal and control of bacteria and other biological contaminants from purified water presents some unique challenges. The ability of certain microorganisms to adapt to and thrive under adverse conditions in marine and freshwater environments provides them with similar advantages in industrial purified water systems (see Chapter 1).

Over the past twenty years, the volume of purified water that has been utilized for various critical manufacturing operations has increased dramatically. Unlike other processing fluids such as cooling waters and metal cutting fluids, purified water is considered a raw material for many manufacturing operations. The quality of numerous manufactured products is directly related to the quality of purified water used in their production. This has been particularly evident in the pharmaceutical and cosmetic industries, where problems related to purified water contamination have resulted in adverse patient reactions and product recalls. The challenge to producers and end-users of purified water is two-fold: to remove contaminants and to prevent subsequent recontamination.

Advances in physical chemistry, electrochemisty, and materials sciences now allow water with a resistivity of 18.24 Mohm·cm to be produced which contains <1 ng/l total organic carbon (TOC) and fewer than 10 abiological particles ≥0.3 um/l. Bacteria and other microorganisms are capable of attachment and growth in such "ultrapure" waters. Consequently, biological fouling problems do arise in these water systems. In many cases, purified water of an otherwise acceptable chemical quality cannot be utilized due to the presence of unacceptable levels of microbial contaminants. New technologies and procedures for identifying, isolating, and treating bacteria in purified waters have trailed those of the abiological contaminants. In this chapter, the sources and consequences of microbial contamination in purified water will be discussed with a common sense approach to preventative maintenance and design strategies for contamination control.

An Overview of Purified Water Systems

With the application of reverse osmosis (RO) systems to industrial water systems in the early 1970's, the production of relatively low-cost purified water became possible. RO systems are designed to remove the bulk of dissolved solids (TDS) from incoming feed water, thus lessening the loading of anion and cation deionizing resins. RO systems have also replaced many diatomaceous earth (D-E) filters, removing bacteria and most organics with molecular weights greater than approximately 300. Because the membranes themselves are subject to biological and abiological fouling problems (see also Chapter 6), some form of pretreatment is usually required. Typically, this consists of a sand-anthracite multi-media filter, nominal filtration to 5 μm, pH adjustment, flocculation, biocide introduction, and ultraviolet irradiation (Figure 1). The recently introduced polyamide thin film composite membranes are relatively intolerant of chlorine and other oxidizing substances; therefore, granular activated carbon (GAC) columns or sodium bisulfite injection are employed when these chemicals are present in the feed water.

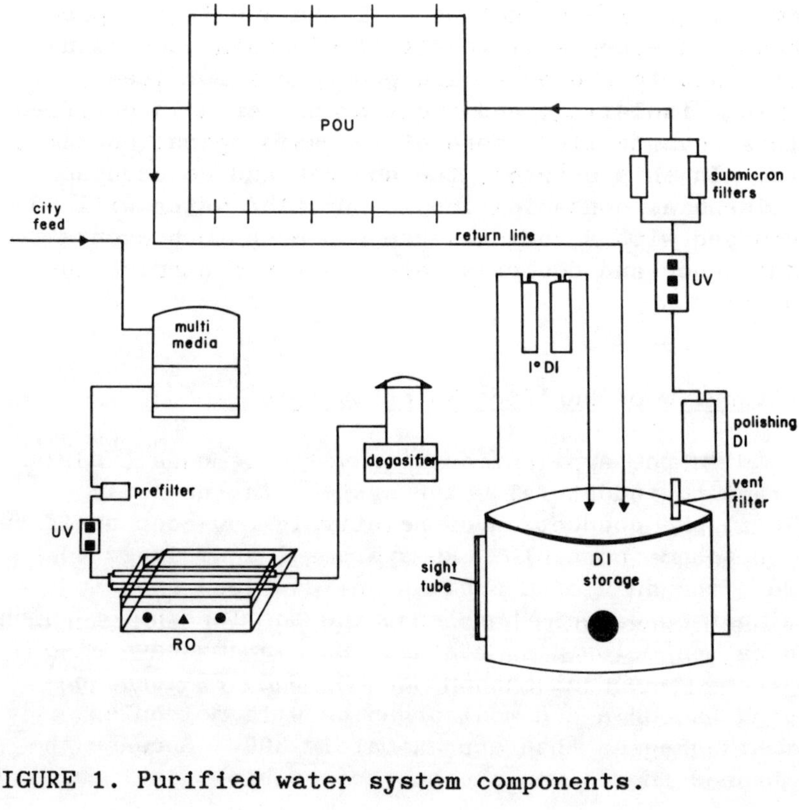

FIGURE 1. Purified water system components.

When acid is required to control carbonate scale formation or to adjust the the feedwater pH, carbon dioxide can form which passes through the membrane and, if not removed, takes up ion exchange capacity on the anion resins. The CO_2 is removed by forced-draft or vacuum-type degasification systems. The more commonly employed forced-draft units pass filtered air through a stream of water over plastic dispersers to displace the more volatile CO_2. Forced-draft degasifiers function much in the same way as a cooling tower; in purified water systems, however, CO_2 is removed instead of heat. Scale inhibitors, such as Floccon 100 (Pfizer), Pretreat 110 (King-Lee Chemical), and Aquafeed-100 (B.F. Goodrich), can be used in some waters to disperse carbonate scales thus precluding the need for acid injection. Typically, these compounds are used in waters with a Langelier Saturation Index of about +2.0 or lower and pH of 8.0 or greater (low CO_2) [29]. The greatest application for these compounds is with polyamide-type membranes, which typically are more pH tolerant than are cellulosic-type membranes.

DI beds contain porous inert spheres, typically 0.5 mm in diameter, which are designed to remove ionized substances from process streams. In some applications, weak-acid type anion resins are used for organic removal; however, other types of DI resins are susceptible to irreversible organic fouling. Typical strong-acid cation resins contain a sulfonic acid functional group in the hydrogen form; strong-base anion resins, a quaternary ammonium functional group in the hydroxyl form. Regeneration of expended DI resins is accomplished by the controlled addition of a strong caustic (anion) or a strong acid (cation) solution, followed by exhaustive rinsing to remove residual regenerant. Larger systems utilize fixed-bed DI units, which are regenerated _in situ_. Many systems also employ exchangeable bottles, containing from 1 to several cubic feet of resin [40]. So-called "nuclear-grade" mixed DI bed units are frequently utilized in power plant condensate polishing loops and in microelectronic applications. The primary purpose of these units is to remove trace levels of silica--typically, they are employed to "polish" deionized water in the range of 12-17 Mohm·cm.

DI or RO waters are stored in fiberglass reinforced plastic (FRP) or stainless steel tanks. Stainless steel tanks are often employed in pharmaceutical applications with "hot loops" for sanitizations (see Design Considerations, below). Corrosion problems associated with ultrapure water (>5 Mohm·cm) often preclude the use of stainless steel tanks in the microelectronic and power industries. Storage tanks are usually vented with a membrane or HEPA filter or a positive pressure nitrogen blanketing system. Level sensing is either via sight tubes or electronic level indicators (accoustical, etc.).

UV and ozone systems are designed to kill microbial populations in the overlying water. Both types of systems are discussed in Chapter 5.

A number of commercially available membrane filter cartridges are suitable for use in purified water. In the pharmaceutical industry, 0.45 μm cartridge filters are standard; in the microelectronic and power industries, 0.22 μm. Membrane filters utilize a combination of mechanical and electrical retention mechanisms to remove particles and microorganisms from purified water. Materials of construction include cellulose esters, Nylon-66, polysulfone, Teflon, polypropylene, and polycarbonate [35]. Cartridge housings contain from 1 to 20 or more 10-60 in. long filters and are constructed of stainless steel, fiberglass, or high-impact plastics.

Distribution systems for purified water are either single-pass or recirculating. Unlike single-pass systems, recirculation returning to the RO or DI storage tanks conserves water and provides for continuous polishing of the product water. Piping systems of 304 or 316L stainless steel are common in pharmaceutical and in some power applications. Sanitary-type, cloverleaf pipe fittings are used in pharmaceutical applications. Extruded piping materials of PVC, polypropylene, PVDF, PFA, and Teflon are commonly employed in ultrapure water systems. With the exception of PVC, these materials can be fitted by butt-welding or threading. Ball-type valves are often selected for use in distribution systems because of their low pressure drops relative to needle valves. However, the relatively large surface area associated with ball valves can be a factor in bacterial generation.

A number of point-of-use systems for the further processing of purified water are available. These include distillation, ultrafiltration, and secondary reverse osmosis. The United States Pharmocopeia [55] recognizes distillation and double-pass RO as acceptable methods for producing pyrogen-free water for injection (WFI). Storage of WFI at 80°C in stainless steel holding tanks is commonly employed in parenteral drug manufacturing operations [41].

Problems Associated With Biological Fouling of Purified Water Systems

Pharmaceuticals, *In Vitro* Diagnostics, and Medical Devices

Microbial contamination of purified water systems has been associated with fouling of dialysis units, RO systems, submicron membrane filters, ultrafilters, water storage facilities, piping, and point-of-use dispensing systems [45,36,16]. Product recalls of a number of contaminated products have been associated with high bioburdens in purified water. A 1975 Canadian recall of 7 cosmetic products [19], a 1984 FDA recall of pH 7.3 calibration buffer ampules [18], and a 1983 FDA recall of *in vitro* diagnostic sodium chloride and purified water solutions [17] are recent examples. Contamination of commercially available distilled water by Gram-negative bacteria has been associated with corneal ulcers of soft contact lens wearers [58]. Inappropriate home preparation of saline lens cleaning solutions with contaminated distilled water was found to be the source of *Pseudomonas* spp. associated with the corneal ulcers.

Anderson *et al*. [2] described contamination problems in commercial distilled water systems used for preparing respiratory therapy solutions. They found that *Psuedomonas pickettii* was capable of passing 0.2 μm membrane filters; 0.1 μm filters were recommended in waters containing high bioburden levels. A number of cases of Legionnaires' Disease were associated with contamination of a hospital distilled water system in Quebec, Canada [27]. Four cases were linked to distilled water used in respiratory therapy equipment; one through a room humidifier fed by distilled water. Following the use of sterile distilled water, no further cases were identified. A study of the USA National Institutes of Health research laboratories revealed that distilled water contained significantly lower bacterial levels than did deionized water [11].

Purified water by USP definition [55] must meet U.S. Public Health Service bacteriological purity regulations for drinking water. Thus, the water must be free from organisms of sanitary significance; typically, a coliform membrane filtration or MPN assay for lactose-fermenting bacteria is performed. Total heterotrophic bacteria assays must also be performed on a frequent basis at all point-of-use locations to insure that levels are "under control" [20]. Although there are no specific limits for total aerobic bacterial levels, most pharamaceutical manufacturing operations maintain total aerobic bacteria levels at less than

10/ml. Water for injection must be sterile and pyrogen-free as defined under current USP guidelines [55]. Rinse waters utilized for certain medical devices, including heart valves, must be pyrogen-free in addition to meeting purified water requirements for bacteriological purity.

Microelectronic Devices

The extremely small feature size of semiconductor devices has created a demand for greater purities of raw materials used in their manufacture. Based upon volume, more purified water is used in the manufacture of these devices than any other material. As particulates, bacteria can cause a conductance phenomena to occur: a short occurs between two adjoining circuit paths, with bacteria acting as the circuit bridge. This particular problem increases in severity as minimum feature sizes approach the 0.1 μm range. Photolithographic deposition processes can also be affected by bacteria acting as particulates. "Shadowing" of oxide layers by bacterial interferences can create electrical anomalies. In sufficiently high numbers, organic contamination by bacteria and their by-products can contribute to the TOC content of purified water.

High TOC levels have been associated with problems in gallium arsenide-based devices. Carbon-containing compounds, such as bacteria, have the potential to act as dopants [42]. The ability of biofilms to concentrate heavy metals and other ionic substances [38] may also be a factor in trace metal contamination of devices. Observed decreases in resistivity across UV systems have been attributed, in part, to the release of ionic materials from damaged bacterial cells. In one study performed at Hewlett Packard Corp. [12], bacteria levels in purified water were positively correlated with device defect frequencies. High bacteria levels have been associated with yield problems in NMOS, CMOS, and bipolar technologies. Biologically mediated failures of back-side metallization adhesion processes have also been reported [9].

Power Generating Systems

Most biological fouling problems in the power industry have been associated with estuarine and freshwater cooling intake systems [15]. Fouling by crustaceans, small fish, and plant materials has been reported [53]. Silica contamination in the condensate polishing loop can result in turbine blade scaling and subsequent failure [14]. Diatoms, which are single-celled plants, are a major source of silica in natural waters. These organisms flourish in warm waters exposed to sunlight. The cell wall of diatoms is composed almost exclusively of silica (silicon dioxide). Microbially influenced corrosion in high temperature purified waters has also been described [52].

Aerospace

Development of the space shuttle (space transportation system) has spurred concern in the area of systems and personnel contamination via fouled water systems. Problems with biological fouling of spacecraft heat exchanger systems [51,32] and potable water systems [48] on the shuttle have been reported. A microbial check valve has been developed for the space shuttle that prevents the transfer of viable microorganisms within the on-board water systems [8]. Detection of bacteria in Apollo spacecraft water systems was performed by measuring pressure drops across membrane filters [49] and the firefly bioluminescent reaction [33]. In the latter assay, ATP (adenosine triphosphate) associated with microbial cells was measured using an enzyme extract from the firefly that produces a light reaction in proportion to the amount of ATP present.

Home Water Treatment

There are an estimated 6 million families using some form of home water treatment device for purifying water in the United States [24]. GAC columns, DI beds, RO systems, and paper filters are utilized in these home systems. Many of these systems are inefficient in removing the organic, halogenated, and particulate contaminants for which they were designed. Contaminated carbon columns, contaminated cellulosic paper filters, and colonized piping surfaces are major sources of bacteria in the treated water. Studies by Geldreich *et al*. [24] demonstrated that some treatment devices allowed passage of challenge bacteria with little or no retention. Colonization was found to be particularly severe during periods of stagnation; e.g., during vacation periods.

The health effects of consuming contaminated drinking water are dependent both upon the dosage and type of microorganism ingested. The World Health Organization [59] and the U.S. National Interim Primary Drinking Water Regulations [56] require that the occurrence of nonfecal coliforms be <3 percent of all samples analyzed over a 12-month period. Some organisms, including *Entamoeba coli*, *Giardia lamblia*, *Shigella* spp., and *Salmonella* spp. can produce illness with a dosage of 100 or less cells (cysts) [1]. Based upon data from a number of studies, home water treatment systems containing carbon filters should not be used on water supplies that do not meet U.S. coliform limits for drinking water.

Factors Influencing Microbial Fouling

As a form of self-replicating particulate, heterotrophic microorganisms utilize various forms of organic carbon as nutrients that promote growth and multiplication in a logarithmic fashion. It is for this reason that it is virtually impossible to maintain any large volume water system in a sterile condition for any length of time (By definition, sterility is the absence of any growing or replicating organisms [3]). As Equation 1 demonstrates, a small number of microorganisms can grow to a large number in a very short period of time.

$$(1) \qquad N = N_i\, 2^{t/g}$$

where,

N = final concentration of microorganisms at time t

N_i = initial concentration of microorganisms ($t=0$)

t = elapsed time in hours

g = doubling time in hours

Heterotrophic bacteria are the predominant microbial contaminants in purified waters. Algae, fungi, and diatoms are sometimes associated with contamination of multimedia sand filters and RO prefilters. However, these organisms are not highly adaptable to the oligotrophic conditions associated with purified waters. In addition, their relatively large size (>5-10 μm) often precludes passage through pretreatment systems. Algae and diatoms can contaminate sight tubes and entry hatch gaskets on storage tanks that are exposed to light. Metabolic by-products associated with these contaminants can then serve as nutrients for bacteria.

Gram-negative, heterotrophic bacteria are the most common type of bacteria present in purified water systems. This group possesses a cell wall composed of multiple layers of peptides, polysaccharides, and phosphorylated compounds. Their cell wall structure provides them with the ability to survive in the extremely hypotonic environment of a deionized water system. Many bacterial genera along with other microorganisms are not capable of maintaining integral cell walls or cell membranes under such extreme conditions. Purified water bacteria are capable of utilizing trace levels of complex organic substances for growth and replication. This same adaptation to oligotrophy (survival in nutrient-limited environments) creates difficulties for culturing these organisms, however. The relatively high nutrient content of microbiological media precludes the growth of many purified water bacteria. There have been some notable improvements in media for stressed environmental organisms such as R2A medium [34]. Saleem and Schlitzer [47] obtained better recoveries of purified water bacteria using membrane filtration and media containing lower concentrations of nutrients. Duplication of *in situ* environmental conditions--particularly in ultrapure waters--is extremely difficult, however.

Microorganisms, like other forms of life, have certain essential requirements for their survival, growth, and replication. Some organisms have unique nutritional or environmental requirements: diatoms, for example, require relatively warm waters exposed to direct sunlight for their proliferation. There are, however, six essential growth promoting substrates which are common to all living organisms. Without a source of carbon, nitrogen, water, sulfur, phosphorous, and trace metals/salts, life cannot exist. It follows, then, that limiting the concentration of one or more of these factors will limit the number of bacteria and other microorganisms in purified water. These growth promoting substrates are listed in Table 1 along with some typical sources.

The major growth limiting substrate found in purified water systems is organic carbon. Heterotrophic bacteria are often able to adapt to varying types

TABLE 1. Growth limiting factors for biological foulants in purified waters.

Growth-Limiting Substrate	Typical Sources
carbon	feed waters (humic and fulvic acids, etc.)
	pipe extractables and lubricants
	microbial by-products
	airborne dust
	treatment chemicals (RO pretreatment)
	personnel (skin flakes, oils, cosmetics)
	membrane filter surfactants
	storage tank entry hatch gaskets
	membrane filter media
water	degasifier HEPA filters
	storage tank sight tubes/vent filters
	dead-legs
nitrogen	feed waters (humic & fulvic acids, nitrates, nitrites, etc.)
	airborne dust
	nitrogen blankets (not a significant source of assimilable nitrogen)
	microbial by-products
phosphorous and sulfur	feed waters (phosphates, sulfates)
	sulfuric acid (RO pretreatment)
	airborne dust
	membrane surfactants
	microbial by-products
trace metals and salts	airborne dust (especially degasifiers)
	piping extractables
	membrane filter extractables
	RO pretreatment chemicals
	metallic system components
light (algae & diatoms)	storage tank sight tubes
	entry hatch gaskets/vent filters

of organics as growth sources. The response to organic carbon loading is dependent upon the concentration and type of organic as well as the population of bacteria present (Table 2). PVC, a commonly used piping material, can be an important source of assimilable carbon for bacteria [4]. In general, organic levels as measured by TOC analysis, for example, are positively correlated with viable bacterial numbers (Figure 2). If the organic species present are not highly labile, a positive correlation between TOC levels and bacterial numbers may not be evident. A major portion of assimilable carbon is converted to cell mass in the form of extracellular polysaccharides as was noted in Chapter 1. These expolymers are most often responsible for the biologically mediated fouling of membrane filters, RO membranes, DI resins, and pipe surfaces.

TABLE 2. Some organics associated with purified water systems

Organic Compound	Typical Source(s)
acetone	reclaim system rinse waters
alkyl phthalates	PVC piping; plasticizers
alkyl sulfonates	reclaim system rinse waters
butylated hydroxytoluene	polypropylene piping; antioxidants
cellulose acetate	membrane filter media; RO membranes
glycerine	reclaim system rinse waters
humic and fulvic acids	surface feed waters
isopropanol	reclaim system rinse waters
stearates	polypropylene piping; lubricants
sulfonated PSDVB	cation resins
thioethers	polypropylene piping; heat stabilizers

Different environments and different seasons of the year bring with them varying conditions for microbial survival. For example, planktonic bacteria levels in most purified water systems tend to increase significantly during spring and early summer as a result of increases in water temperature. Changes in feedwater nutrient levels are often tied to the seasons as well. During spring runoff conditions, surface waters typically contain increased concentrations of carbon-containing plant degradative products such as the humic and fulvic acids. Municipal water purveyors switch sources depending on economic and quality factors. In southern California, for example, most water is received from the Owens Valley, the Colorado River, surface well waters, and mixtures of all three. Within a 24 h period, TDS and TOC levels can vary from 200-700 ppm and 1-50 ppm, respectively. Keeven _et_ _al_. [28] found that decreased oxygen levels in high purity waters increased the survival rates of _Pseudomonas cepacia_, _Pseudomonas putida_, _Pseudomonas aeruginosa_, and _Enterobacter gergoviae_.

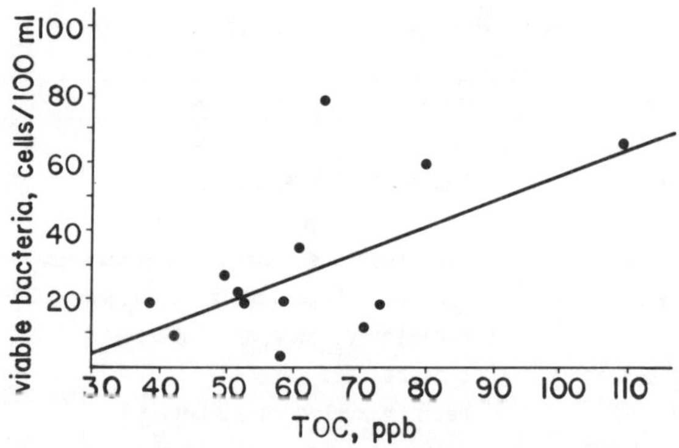

FIGURE 2. Relationship between total oxidizable carbon (TOC) and viable bacteria levels in an 18 Mohm·cm water system. Correlation coefficient=0.594 ($P<0.05$)

Routine monitoring of both abiological and biological contaminant levels in feed waters for pretreatment is essential to the maintenance of systems operations.

Finally, hydraulic parameters can also influence microbial fouling. The presence of dead-leg areas in purified water systems is associated with a concentration polarization phenomenon. Water is an excellent solvent for many materials: high purity water can be considered an aggressive fluid for stainless steel, certain plastic piping materials, some membrane filter media, and other materials of construction. Extractable substances accumulate in dead-leg areas serving both as a source of nutrients for bacteria as well as contaminants for various manufactured products.

Design and Preventative Maintenance Considerations

Granular Activated Carbon (GAC) Beds

GAC systems are employed for the removal of organics and halogenated compounds (chlorine, chloramines, iodine, etc.). Before the advent of industrial RO systems in the early 1970's, GAC units were the major organic removal component on purified water systems. Since the development of cellulose acetate RO membranes, GAC units have sometimes been installed downstream of the RO to remove halogenated compounds. Chlorine and other oxidizing agents are poorly rejected by RO membranes and thus may contact DI resins, resulting in resin degradation. Thin film composite polyamide RO membranes are also sensitive to low levels of halogenated compounds; GAC beds are often employed upstream of these membranes as a protective measure. Chloramines present a particular problem for renal dialysis systems, since, without appropriate pretreatment, chloramines can pass the dialysis membranes into the dialysate. GAC columns are often used for chloramine removal upstream of dialysate RO/UF systems.

The problem with GAC beds used in these and other types of systems is that they generate *in situ* large bacterial populations [57,16]. The large surface area of the carbon fines provides ample space for bacterial attachment, biofilm formation, and endotoxin production. Fernandez et al. [22] found effluents from GAC filters contained from 10 to 1000 times more bacteria than the city feed water. Additionally, the organic and inorganic materials trapped by the GAC provide an excellent nutrient source for biofilms.

Where possible, GAC beds should be eliminated and alternative treatments employed. In many systems, sodium bisulfite can be used in place of GAC beds for the reduction of halogenated compounds. Although the injection of sodium bisulfite results in an increase in feedwater TDS--typically 10-15 ppm--the benefits of its application outway the relatively small increased cost of TDS removal. An added benefit of sodium bisulfite is its biocidal activity, which may improve the bacteriological quality of RO feed and permeate waters. Sufficient sodium bisulfite to neutralize 10 ppm residual chlorine appears to be somewhat effective, although RO and UF membrane manufacturers should be contacted regarding maximum tolerable limits.

Le Chevallier et al. [31] found that bacteria associated with GAC fines were extremely resistant to chlorine. The chlorine resistance was mediated by biofilm production coupled with the protection afforded by fissures in the fines. Because of the large organic demand associated with GAC beds, chlorine sanitizations are often ineffective. Steam or hot water sanitizations are effective; however, the organic removal capacity is reduced by the high temperatures required. Collentro [7] recommends caustic regeneration as an effective and safe sanitization/ regeneration process.

Degasifiers

A diagram of a typical forced-draft degasifier is shown in Figure 3. The major areas of contaminant introduction for these units are the air intake filter, the clear well, and the chimney exhaust. Placement of the degasifier upstream of the RO unit can reduce permeate contamination; however, larger volumes of acid are typically required for CO_2 removal in this configuration. Since large volumes of ambient air are passed through the intake filter, it is essential that it be properly maintained. A HEPA-type filter should be used with a particulate removal efficiency of 99.997% or better. The use of a wide mesh, furnace-type prefilter will prolong the life of the more expensive HEPA filter. The prefilter should be replaced at six month intervals; more frequently if

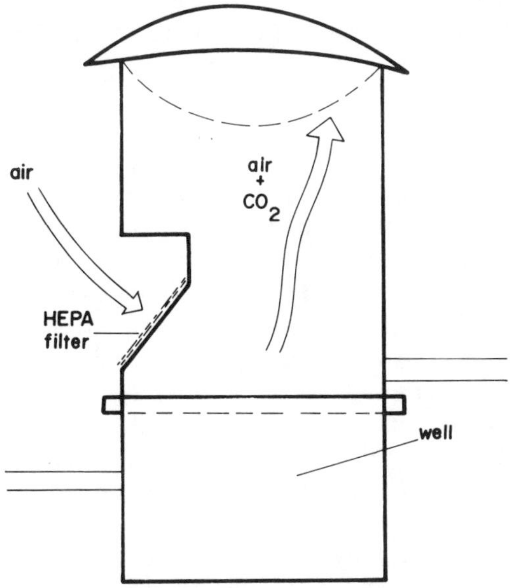

FIGURE 3. Forced-draft degasifier components.

the system is located in a particulary dirty area. If filters become wetted, they can wick contaminants into the clear well along with creating a large pressure drop. Wet prefilters and HEPA filters should be replaced immediately.

In some systems, the degasifier is shut down when the RO is off. In the absense of positive pressure, contaminants--including dirt, insects and birds--can enter open exhaust vents. A screen and protective flap should be installed to prevent introduction of contaminants. Dirt and macroparticulates can accumulate in the bottom of the clear well. This section should be opened for inspection on a monthly basis and cleaned as needed (see Treatment, below). Fouling of plastic packing media can reduce CO_2 removal efficiency and act as a source of organics and bacterial contaminants. It is often less costly to replace fouled packing material than it is to clean.

Storage Tanks

Biological fouling problems occur in large and small-volume RO and DI water storage tanks. Forced-draft degasifiers and contaminated DI beds are the major sources of microbial and organic contamination for storage tanks. Faulty entry-hatch gaskets, which are a source of organic carbon, can admit light and dirt into tanks. These gaskets should be replaced each time the entry-hatch is opened for inspection, at least on a twice-yearly basis. Sight tubes should be replaced with electronic level indicators. Algae and large numbers of bacteria can often be isolated from sight tubes.

Vent filters allow the tank to "breathe", preventing aspiration of contaminants from the ambient atmosphere. These filters should be constructed of a hydrophobic material (Teflon, etc.) and oriented in a vertical position to prevent residual water accumulation. Cellulosic-type vent filters have been known to become saturated with water, preventing displacement of the overlying air in the tank. This has resulted in implosions of tank walls in the presence of relatively low ΔP's (≈ 4 psig) created by normal demand. Wetted vent filters are also a dead-leg area, prone to bacterial and algal contamination. A filter with an absolute pore size rating of 0.45 μm or a HEPA filter provide adequate flow and contaminant protection.

Storage of RO water is sometimes preferred over DI water storage when a small chlorine or iodine residual is maintained in the tank for bacteria control. Typically 0.2-0.4 ppm residual chlorine and 0.4-1.0 ppm residual iodine are maintained via RO feed injection (Figure 4). However, this can present a

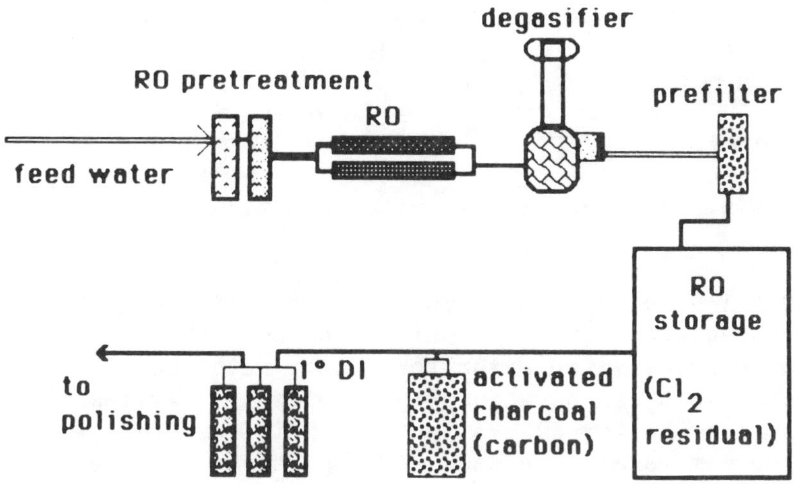

FIGURE 4. RO water storage with a GAC column for chlorine removal.

problem for DI resins. Problems associated with the use of GAC beds for halogen removal were discussed above. The storage of DI water with continuous ozonation provides one alternative to this situation (Figure 5). Provided the FRP and attendant piping materials of construction are compatible with ozone,

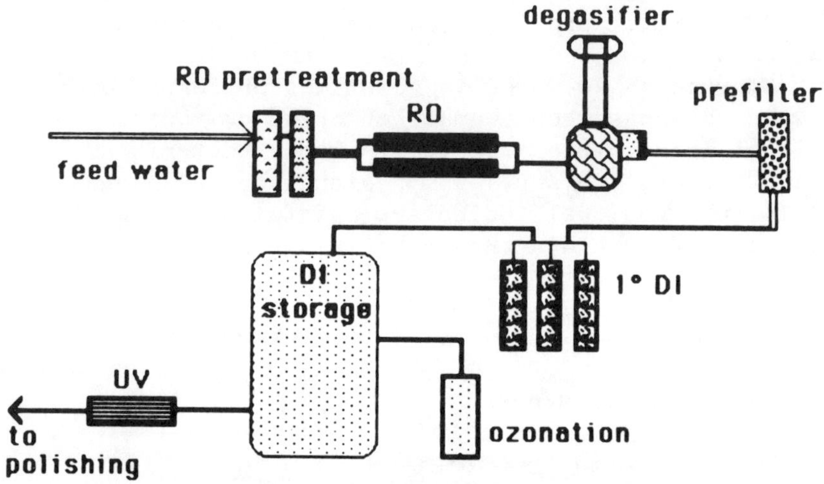

FIGURE 5. DI water storage with ozonation.

an 0.5-1.0 ppm residual ozone concentration will provide adequate bacterial control in the tank. Ozone must be inactivated prior to contact with the DI resins: this can be accomplished with a UV system (see also Chapter 5). A "hot loop" is sometimes employed with stainless steel storage tanks. Bacteria control is obtained when the tank water is maintained at 80°C by recirculation through a heat exchanger [25].

Tank water turnover rates are another important consideration in biological fouling control. In tanks with turnover rates of <1/48 h, bacteria levels can increase without some form of continuous treatment. Storage tanks should not, therefore, be grossly oversized for the intended usage. Such tanks can be treated on a continous basis, however, with ozone or heat as mentioned above. Another alternative is recirculation through a UV purifier system. A 5,000 gallon storage tank, for example, would require a UV system sized for 3.5 gpm in order to effect a turnover rate of 1/24 h. This type of design is particularly well suited to systems that are shut down for frequent periods of time (night shifts, weekends, holidays, etc.). Return (and recirculation) lines should always be plumbed to the top of tanks to provide adequate mixing.

DI Systems

Both fixed bed systems and exchangeable DI bottles have a potential to encourage bacterial growth. As with GAC fines, DI resins contain relatively large surface areas for bacterial attachment. Fortunately, regeneration with either acid or caustic regenerants effectively sterilizes the resins. However, regeneration is performed on the basis of decreases in resistivity or increases in silica content, not bacterial breakthrough. Ion meters (resistivity, silica, etc.) will not detect the presence of bacteria. Users of exchangeable bottles should take care to assure that the resins are not allowed to remain on "stand-by" for extended periods of time. DI resin bottles are delivered wet, and, on standing in the sun, can promote the growth of large numbers of bacteria in a matter of a few days. Continuous recirculation through the DI beds at the highest possible hydraulic loading with effective pretreatment (UV, prefiltration) will aid in limiting nutrient loading and subsequent fouling. Rinsing of regenerants from resin beds should always be performed using RO or DI water. Some users attempt to economize by using city feed water as a rinsing agent. Not only is ion exchange capacity wasted by this practice, but large numbers of organic, microbial, and particulate contaminants are introduced into the system.

Polishing resins are a major source of bacteria since they often remain on-line for extended periods of time. These resins usually receive a relatively small amount of ionics in their feedwater. Since regeneration is performed on the basis of ionic breakthrough, bacterial fouling is often undetected. These resins, in particular, should be regenerated on the basis of significant increases in bacterial numbers rather than on ion breakthrough.

Distribution Systems

Dead-legs and inadequate flow rates account for most of the bacterial contamination problems observed in distribution and point-of-use dispensing systems. A dead-leg may be defined as any length of piping with no outlet that is not open to constant recirculation (Figure 6). Along with acting as an area of concentration for growth promoting factors, dead-legs impede the passage of treatment chemicals into areas where they are most needed. Rinsing of contaminants and

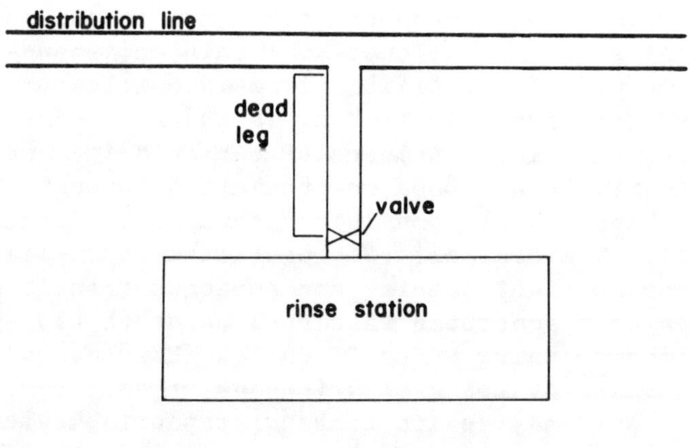

FIGURE 6. Dead-leg rinse station.

treatment chemicals becomes difficult: contaminants and residual treatment chemicals eventually contaminate otherwise clean areas of the distribution loop. Some designers consider a dead-leg to be equivalent to a length of piping greater than 4-6 times its diameter. Continuous, high velocity flow through distribution system is essential for controlling biofouling problems. Linear fluid velocities in the range of 5-10 ft/sec are recommended by a number of designers [39,46]. Increased fluid velocities increase the frequency of microbial inactivation and removal via UV and filtration systems, respectively. Bacterial attachment can occur in the presence of relatively large Reynold's numbers and high fluid velocities, however (see also Chapter 2). Fluid velocity may be calculated by the formula shown in Equation 2.

$$\text{(2) Velocity, ft/sec} = \frac{\text{flow rate, gpm} \times 144\ \text{in}^2/\text{ft}^2}{\pi \times (0.5 \times \text{pipe ID, in})^2 \times 7.48\ \text{gal/ft}^3 \times 60\ \text{sec/min}}$$

Distribution piping should always be plumbed as closely as is possible to point-of-use dispensing areas. The frequently observed design shown in Figure 6 has been associated with bacteria and endotoxin contamination in both medical and non-medical purified water systems [13]. Dead-leg designs have been identified as one of the major contributing factors to bacterial and endotoxin contamination of dialysate delivery systems [10]. A more appropriate design for a point-of-use distribution loop is shown in Figure 7. This type of design can be accomodated by the use of backpressure regulating valves to maintain loop pressure. Other design criteria for maintaining good flow rates and acceptable pressure drops without the incorporation of dead-legs in large distribution systems are discussed by Nickerson [39] and Bukay [5]. Recirculating hand sprayers should also be considered for point-of-use dispensing systems (Figure 8).

A barrier faucet was found to offer some protection from airborne contaminants in hospital distilled water systems. However, once one outlet became contaminated, the entire distilled water system was subsequently fouled [54]. A UV sterilizer faucet for controlling airborne contamination of purified water

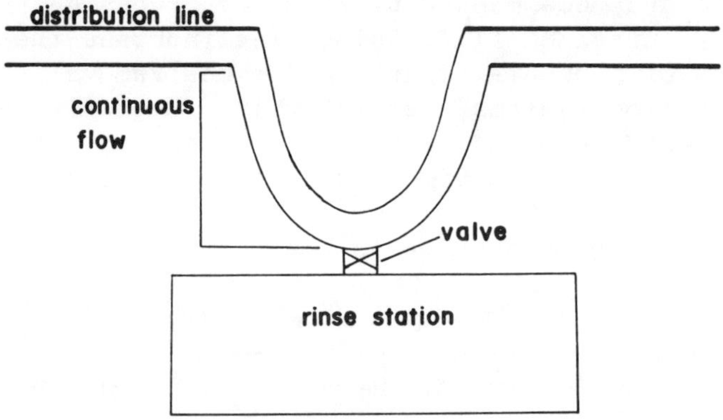

FIGURE 7. Continuous flow point-of-use dispensing system.

systems at point-of-use locations has been developed for use in microelectronic, research laboratory, and hospital applications [23]. The unit consists of a control module (power source) and a UV faucet head containing a 254 nm UV lamp. The role of airborne bacteria in contaminating water systems via point-of-use locations has not been well established, however. Further studies on this topic and the efficacy of the UV faucet are needed.

Threaded and socket-type connections should be avoided whenever possible. In pharmaceutical applications, butt-welded 804L or 816L stainless steel piping with sanitary "cloverleaf" fittings is often recommended [44]. The FDA has published standards related to acceptable distribution system materials of construction [21]. Butt-welding should be used with plastic piping materials whenever possible. This type of weld is free from tortuous pathways where microorganisms and other particulates can concentrate. Threaded fittings and rough welds afford microbial contaminants

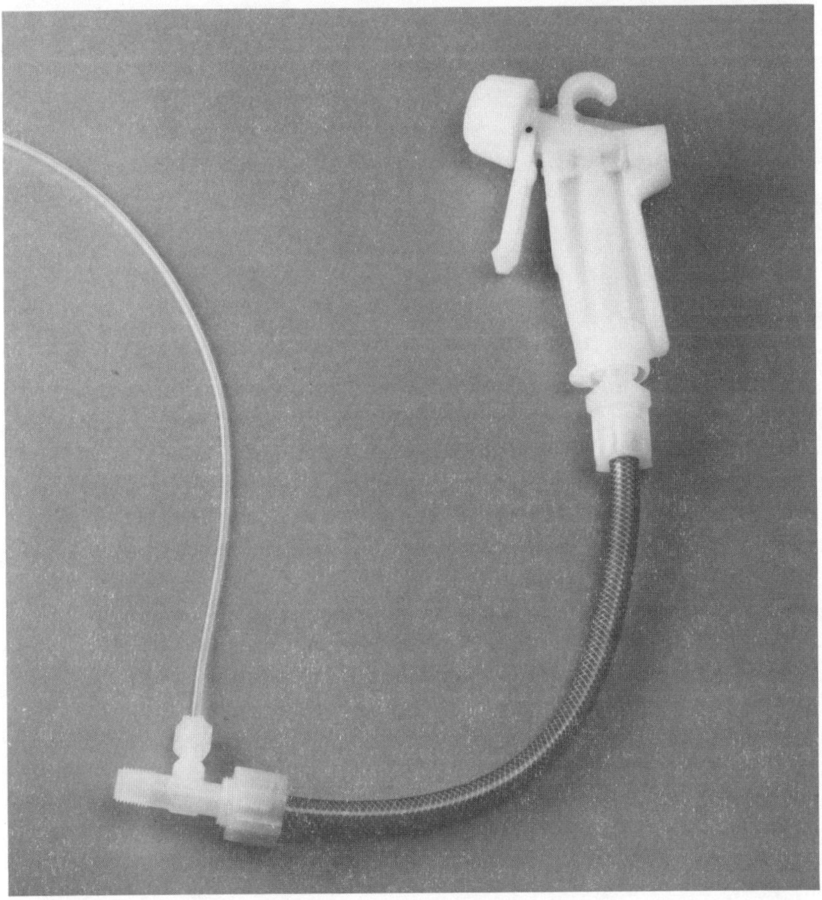

FIGURE 8. Recirculating hand-spray rinser for point-of-use dispensing of purified water (photograph courtesy of Fluoroware, Inc.).

partial protection from the flushing effect of turbulent flow and/or inactivation by treatment chemicals. Improperly glued or threaded PVC piping provides an ideal reservoir for bacterial contaminants (Figure 9).

FIGURE 9. Schedule 80 PVC pipping threaded connection. Part of the threaded connection has become disengaged and is in contact with the water stream.

Membrane Filters

Contamination of submicron pore-sized membrane filters is the most recognizable consequence of biological fouling. In sufficient numbers, bacteria can plug the feedwater side of membranes, growth through the membrane media, and contaminate the product water side with whole cells and cell fragments. Contamination is usually evidenced by an increase in filter ΔP and discoloraton of feedwater surfaces (Figure 10). Bacterial grow-through has been documented by a number of workers [50,26,6]. Grow-through appears to be a time-dependent process. In many pharmaceutical process applications, for example, filters are utilized during an 8-12 h fill period, then discarded. Grow-through problems in these types of applications appear to be rare. A "blow-through" phenomenon has also been described by Christian and Meltzer [6], whereby filter shock caused by rapid increases in ΔP resulted in bacterial passage.

Proper control of the feedwater bacterial populations is the only way to insure membrane filter integrity. Replacement of filter cartridges during system cleanings is essential--in situ sanitization of used filters is not efficacious and may lead to chemical degradation of the filter media. A study by Leahy and Gabler [30] indicated that hydrophobic vent filters retarded bacterial grow-through when compared with hydrophilic-type filters. One explanation for the observed difference may be the smaller volume of available water for bacterial growth in the hydro-

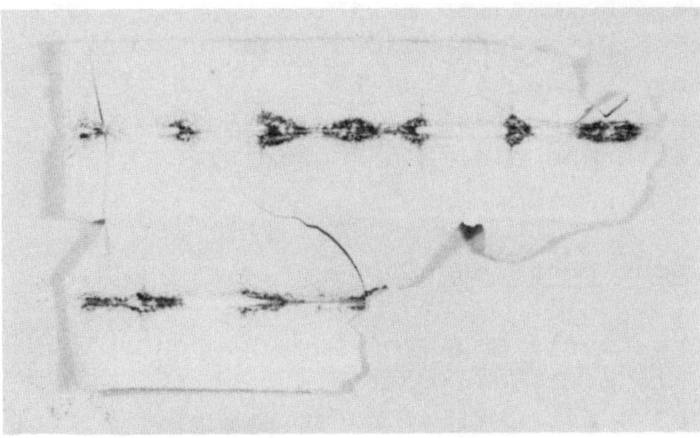

FIGURE 10. Cellulose acetate membrane media contaminated by bacteria. Membrane material is from an 0.2 μm, 10 in. cartridge filter that exhibited a ΔP of 30 psig.

phobic media. Rechen [43] has noted the benefits of ultrafiltration at point-of-use locations. The tangential flow associated with most UF units allows bacteria and other particulates to be swept from the membrane surface, thus precluding grow-through. In addition, UF units with molecular weight cutoffs of 10,000 effectively remove pyrogens from feed waters.

The use of ΔP measurements as a specification for filter replacements is not recommended. Bacterial grow-through or blow-through can occur long before measureable increases in filter ΔP are observed.

Rather, viable and/or total counts of bacteria in filter effluents should be used as a specification for filter replacement. Filter housings should be inspected and sanitized each time filters are replaced. Mechanical scrubbing or hydraulic treatment with an appropriate cleaning compound is recommended (10% (v/v) hydrogen peroxide works well for this application). Housings should also be inspected for any sign of corrosion or other structural defects. Filters should always be handled aseptically with sterile gloves--never with bare hands. Bubble-point or forward flow diffusion testing is required of all pharmaceutical aseptic process filters prior to use to validate manufacturers claims of porocity.

Treatment Considerations

Any treatments employed for the control of microbial foulants must be compatible with the water system materials of construction. As an example, sodium hypochlorite should never be used for routine sanitizations of stainless steel storage tanks or distributions systems. High concentrations of chloride ions can cause chloride stress cracking, a form of corrosion, in these types of systems. Disposal of treatment wastes and rinse waters is another important consideration. Formaldehyde, for example, is widely used in the pharmaceutical industries as a distribution system cleaner. The strict environmental disposal requirements in many municipalities have made its use economically prohibitive. Finally, treatment efficacy must be assessed for each application. Both planktonic and sessile microbial populations must be inactivated and removed from the overlying water and surfaces. Generally, planktonic bacteria are more easily inactivated than are attached populations. Figure 11 illustrates the efficacy of a quaternary ammonium compound against these two populations. The subject of treatment is discussed further in Chapter 9.

The treatments typically applied to purified water storage and distribution systems are shown in Table 3. On a mg/l basis, the effectiveness of chemical treatments may be listed in the order of ozone>sodium hypochlorite>iodine>quaternary ammonium compounds>hydrogen peroxide>formaldehyde ≈ glutaraldehyde>ionic surfactants>nonionic surfactants [37]. While continuous ozonation at levels of 0.5-1.0 ppm appears to be effective in controlling fouling, slug dosages applied to heavily fouled systems are not highly effective.

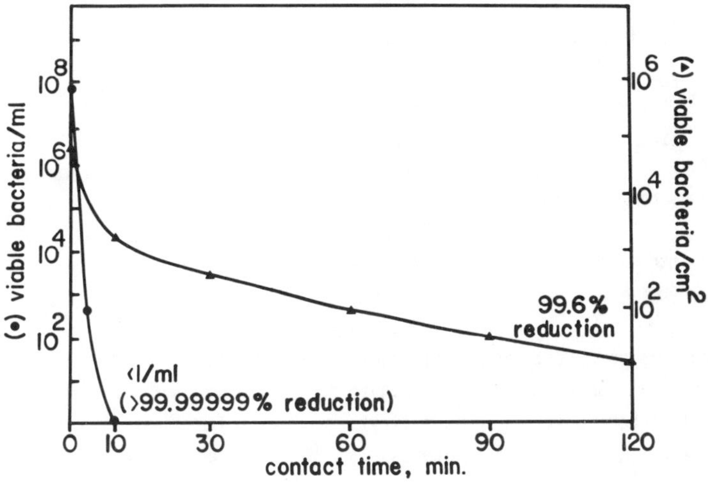

FIGURE 11. Effect of a quaternary ammonium cleaner on sessile and planktonic *Pseudomonas cepacia* populations in purified water.

Mechanical or hydraulic cleaning of storage tank

surfaces is required for effective biofilm removal. Youngberg [60] recommends installation of an integral, high pressure tank cleaning nozzle (Spraying Systems Co., Model 12900-1) at the top of storage tanks. If corrosive chemicals are to be used in treating the tank, the nozzle should be coated with a protective material such as Teflon (DuPont 945-102, Blue). Appropriate safety procedures should always be followed when confined spaces are entered. These include the stationing of at least one person outside the tank supervising work inside, use of oxygen deficiency meters, self-contained breathing apparatus, ventilation fans, and first aid supplies. Nitrogen blanketing systems should always be locked "off" when personnel are working inside the tank.

TABLE 3. Purified water system treatments

Treatment	Dosage Regime	Treatment Duration
chlorine	50 ppm	1-2 h
quaternary ammonium compounds	300-1000 ppm	2-3 h
iodine	50-100 ppm	1-2 h
hydrogen peroxide	10% (v/v)	2-3 h
formaldehyde	1% (v/v)	2-3 h
ozone	0.5-1.0 ppm	continuous
high pressure spray (surfaces)	as required	as required
steam (GAC beds only)	as required	as required
caustic (GAC beds only)	5% (v/v)	1 h
hot water	80°C	continuous

Test Methods

Effective monitoring of contaminants in purified waters is essential to the effectiveness of an overall preventative maintenance program. Bacterial levels, in particular, are an excellent predictor of future fouling potential. Test methods for bacteria must take into account the unique environment of a purified water system. Table 4 summarizes test methods appropriate for purified water microbial assays. Although there are a number of other methods available for microbial analysis in water systems, they do not provide adequate sensitivity or precision for purified water systems. For example, ATP assays are widely employed for enumerating microbial populations in petroleum applications. However, the assay provides only a rough estimate of microbial numbers in excess of 10,000 cells/ml. Future enumeration methods will utilize cellular markers such as fatty acid content as indirect measures of low numbers of microorganisms in purified waters.

TABLE 4. Bacterial enumeration methods for purified waters

Type of Test	Analytical Technique	Analytical Sensitivity
Viable Count	standard pour plate	100 cells/100 ml
	membrane filtration	<1 cell/100 ml
Direct Count	epifluorescence microscopy	<10 cells/100 ml[1]
	scanning electron microscopy	<10 cells/100 ml[1]
Biochemical	fatty acid (PLFA)-GC/MS	<100 cells/100 ml[1]

[1]sensitivity is dependent upon sample volume

Membrane filtration and direct-count epifluorescence microscopy are both acceptable enumeration techniques (see also Chapter 10). Novel media for enumerating bacteria in purified waters include R2A, 0.3% TSA, and Pseudomonas Selective Isolation Agar (PSIA). This medium appears to be superior to Cetrimide, Pseudosel, and Dettol agars for isolating *Pseudomonas* spp. Incubation at 30°C for 72 h followed by 96 h at room temperature is recommended for this medium. Sampling techniques for microbial enumeration are also discussed in Chapter 10.

Bioburden identification is required in pharmaceutical and medical device applications. For example, Pseudomonas cepacia, a common purified water isolate, is of concern to eye-care product manufacturers (treatment and cosmetic). Identification tests for this organism are necessary in these applications. Identification of purified water isolates is useful in determining bioburden sources in contaminated in-process and finished products. Selected colonies are subjected to various cultural and biochemical assays discussed in Chapter 11. The utility of identifying microbial contaminants in non-medical purified water systems is often questionable. Generally, differentiating macrofoulants such as crustaceans from microbial foulants is adequate for the selection of appropriate treatments. Identification of fouling populations should only be considered when treatments appear to be ineffective and resistance is suspected.

References

1. Akin, E.W. 1982. Infective dose of waterborne pathogens. Proceedings of the 2nd national USEPA symposium on municipal wastewater disinfection, Orlando, FL.

2. Anderson, R.L., L.A. Bland, M.S. Favero, M.M. McNeil, B.J. Davis, D.C. Mackel, and C.R. Gravelle. 1985. Factors associated with Pseudomonas pickettii intrinsic contamination of commercial respiratory therapy solutions marketed as sterile. Appl. Environ. Microbiol. 50:1343-1348.

3. Block, S.S. 1983. Disinfection, steriliza-

tion, and preservation. Lea & Febiger, Philadelphia.

4. Booth, G.H., A.W. Cooper, and J.A. Robb. 1968. Bacterial degradation of plasticized PVC. J. Appl. Bacteriol. 31:305-310.

5. Bukay, M. 1986. Dead legs: a widespread threat to DI water system. Ultrapure Water 4(3):70.

6. Christian, D.A. and T.H. Meltzer. 1986. The penetration of membranes by organism grow-through and its related problems. Ultrapure Water 3(3):39-44.

7. Collentro, W.V. 1986. Pretreatment part II: activated carbon filtration. Ultrapure Water 2(6):45-48.

8. Colombo, G.V., D.F. Putnam, and R.L. Sauer. 1978. Microbial check valve for shuttle. ASME intersociety conference on environmental systems, July, 1978, San Diego, CA.

9. Craven, R.A., A.J. Ackermann, and P.L. Tremont. 1986. High purity water technology for silicon-wafer cleaning technology. Microcontamination 4(11):14-21.

10. Dawids, S.G. and R. Vejlsgaard. 1976. Bacteriological and clinical evaluation of different dialysate delivery systems. Acta. Med. Scand. 199(3):151-155.

11. Deroos, R., D. Vesley, A. Duchene, and L.J. Hart. 1976. High purity water supplies for biomedical research laboratories. Health Lab. Sci. 13(1):11-19.

12. Dial, F. and T. Chu. 1987. The effect of high bacteria levels with low TOC levels on bipolar transistors: a case study. p. 178-193. Proceedings of the 6th annual semiconductor pure water conference, January, San Jose, CA.

13. DuMoulin, G.C., E.C. Coleman, and J. Hedley-Whyte. 1987. Bacterial colonization and endotoxin content of a new renal dialysis water system composed of acrylonitrile butadiene styrene. Appl. Environ. Microbiol. 53:1322-1326.

14. Electric Power Research Institute Technical Report RP2174-09. 1983. Reconstruction of pH histories of 10 regional integrated lake-watershed acidification study (RILWAS) lakes. EPRI, Palo Alto, CA.

15. Electric Power Research Institute Technical Report CS-3550. 1984. Condenser macrofouling control technologies. EPRI, Palo Alto, CA.

16. Favero, M.S., J.N. Petersen, L.A. Carson, W.W. Bond, and S.H. Hindman. 1975. Gram-negative water bacteria in hemolysis systems. Health Lab. Sci. 12:321-334.

17. FDA Recalls & Court Actions, August 31 & September 7, 1983. 1983. The Gray Sheet 9(37):I&W-9-11.

18. Recalls & Court Actions, November 7 & 14, 1984. 1984. The Gray Sheet 10(47):I&W-6,7.

19. FDC Reports. 1975. The Pink Sheet 37(8):T&G-6.

20. FDC Reports. 1984. FDA inspection program emphasizing purified water system validation. The Pink Sheet 46(43):T&G-1,2.

21. Federal Register June 1, 1976. 41FR:22211-22213.

22. Fernandez, V., R.L. Calderon, and P.H. Hauser. 1986. Colonization of GAC water filters by water distribution bacteria. Abstracts of the 1986 American Society for Microbiology Annual Meeting, March, Washington, D.C.

23. Galtek Corp. Technical Note. Epco UV sterile faucet system. Galtek Corp., Chaska, MN.

24. Geldreich, E.E., R.H. Taylor, J.C. Blannon and D.J. Reasoner. 1985. Bacterial colonization of point-of-use water treatment devices. J. Am. Water Works Assoc. 77:72-80.

25. Giorgio, R.J. 1978. Considerations in the design of hot circulating water-for-injection systems. Pharm. Technol. 2(2):19-24.

26. Howard, G. and R. Duberstein. 1980. A case of penetration of 0.2 um-rated membrane filters by bacteria. J. Parent. Drug Assoc. 34(2):95-102.

27. Joly, J.R., P. Dery, L. Gauvreau, L. Cote, and C. Trepanier. 1986. Legionnaires' Disease caused by <u>Legionella-dumoffii</u> in distilled water. Can. Med. Assoc. J. 135(11):1274-1277.

28. Keeven, J., J.V. Sorrentino, and B.T. DeCicco. 1984. Nutritional and physical factors influencing growth of bacteria in high purity waters. Abstracts of the 1984 American Society for Microbiology Annual Meeting, March, St. Louis, MO.

29. Kraft, R.L. 1985. Elimination of degasifiers

from high purity water systems. Ultrapure Water 2(2):20-27.

30. Leahy, T.J. and F.R. Gabler. 1984. Sterile filtration of gases by membrane filters. Biotechnol. and Bioengin. 26:836-843.

31. LeChevallier, M.W., T.S. Hasenauer, A.K. Camper, and G.A. McFeters. 1984. Disinfection of bacteria attached to granular activated carbon. Appl. Environ. Microbiol. 48:918-923.

32. Lechtman, M.D. and R. Nachum. 1968. Microbiological evaluation of a typical suit loop heat exhanger during manned testing. Aerospace Med. 39:988-993.

33. Levin, G.V., A.R. Slonim, and E. Usdin. 1968. Rapid detection of microorganisms in aerospace water systems. Aerospace Med. 39:14-16.

34. Meade, E.H. and P. Song. 1985. Bacteria recovery in high purity water systems. Ultrapure Water 2(6):35-39.

35. Meltzer, T.H. 1987. Filter and filter holder-housing design. p. 772-780. In T.H. Meltzer (ed.), Filtration in the pharmaceutical industry. Marcel Dekker, New York.

36. Mittelman, M.W. 1985. Biological fouling of purified water systems, part 1, bacterial growth and replication. Microcontamination 3(10):51-55;70.

37. Mittelman, M.W. 1986. Biological fouling of purified water systems: part III, treatment. Microcontamination 4(1):30-40;70.

38. Mittelman, M.W. and G.G. Geesey. 1985. Copper-binding characteristics of exopolymers from a freshwater-sediment bacterium. Appl. Environ. Microbiol. 49:846-851.

39. Nickerson, G.T. 1986. DI water distribution loop design. Ultrapure Water 3(4):52-55.

40. Owens, D. 1985. Equipment for ion exchange. Ultrapure Water 2(2):12-17.

41. Parenteral Drug Association. 1983. Design concepts for the validation of a water for injection system, Technical Report No. 4. Parenteral Drug Association, Philadelphia.

42. Poirier, S.J. 1985. The new role of TOC analysis in pure water system management. p. 197-210. Proceedings of the 4th annual semiconductor pure water conference, January, San Francisco, CA.

43. Rechen, H.C. 1984. The role of ultrafiltration in pyrogen-free purified water. Pharm. Manufactur. 1(7):28-31;55.

44. Rechen, H.C. and R.T. Miller. 1983. Design considerations for a pharmaceutical-grade RO-DI water purification system. Particulate & Microbial Control September/October, p. 30-34.

45. Ridgway, H.F., A. Kelley, C. Justice, and B.H. Olsen. 1983. Microbial fouling of reverse osmosis membranes used in advanced wastewater treatment technology. Appl. Environ. Microbiol. 45:1066-1084.

46. Riley, C.D. 1983. Synergetic design of ultrapure water systems for microelectronic

applications. Microelectronic Manufacturing & Testing July, p. 21-23.

47. Saleem, M. and R.L. Schlitzer. 1983. Comparative recovery of bacteria from purified water by the membrane filtration technique and the standard plate count method. Abstracts of the 1983 American Society for Microbiology Annual Meeting, March, New Orleans, LA.

48. Sauer, R.L. 1981. The potable water. NASA STS-1 Technical Report No. N82-15711 06-51, p. 63-66.

49. Sharpley, J.M. 1967. Detection of microbial contaminants in space system water supplies. USAF Report No. AD-687853;AMRL-TR-68-67.

50. Simonetti, J.A. and H.G. Schroeder. 1984. Evaluation of bacterial grow-through. J. Environ. Sci. 27(6):27-32.

51. Sofios, M. 1973. Microbial contamination of water--traditional and space-age problems and approaches. SAE, ASME, AIAA, ASMA, and AIChE intersociety conference on environmental systems, July 1973, San Diego, CA.

52. Stoecker, J.G. and D.H. Pope. 1986. Study of biological corrosion in high temperature demineralized water. Paper no. 126, Proceedings of the National Association of Corrosion Engineers annual meeting, NACE Publications, Houston, TX.

53. Strauss, S.D. and P.R. Puckorius. 1984. Cooling-water treatment for control of scaling, fouling, water corrosion. Power 128(6):1-24.

54. Takata, K., O. Fujishita, H. Kameya, S. Ikemura, and M. Uehara. 1979. Microbial barrier faucets for sterile solutions 2. Trial manufacture of microbial barrier faucets. Chem. Pharm. Bull. (Tokyo) 27:1647-1662.

55. United States Pharmacopoeial Convention. 1980. The United States Pharmacopoeia, XXI. Mack, Philadelphia.

56. U.S. Environmental Protection Agency. 1976. National interim primary drinking water regulations. EPA-570/9-76-003.

57. Weber, W.J., M. Pirbazari, and G.L. Melson. 1978. Biological growth on activated carbon: an investigation by scanning microscopy. Environ. Sci. Technol. 12:817-819.

58. Wilson, L.A., R.L. Schlitzer, and D.G. Ahearn. 1981. *Pseudomonas* corneal ulcers associated with soft contact lens wear. Am. J. Ophthalmol. 92(4):546-554.

59. World Health Organization. 1984. 1984 guidelines for drinking water quality, Vol. 1. Recommendations. WHO, Geneva, Switzerland.

60. Youngberg, D.A. 1985. Sterilizing storage tanks in a pure water system. Ultrapure Water 2(4):45.

CHAPTER 8. EFFECTS OF FOULING BIOFILMS ON SYSTEMS OPERATIONS

W.F. McCoy, Great Lakes Chemical Corp., Lafayette, IN

The accumulation of unwanted materials onto engineered surfaces is generally referred to as fouling. There are three major categories of industrial fouling due to biofilm formation: a) product quality degradation (Chapters 4,6, and 7), b) corrosion (Chapter 3), and c) energy losses. Biofouling processes that lead to industrial energy losses are usually due to the effect of biofilms on moving fluids.

Fouling deposits cause significant energy losses because of heat transfer resistance and fluid frictional resistance. Fouling biofilms affect fluid dynamics in conduits (e.g., consider heat exchanger tubes or drinking water distribution pipes). Biofilms are fully hydrated deposits (>98% water) and thus they can significantly decrease conductive heat transfer [5-7].

The overall annual cost due to heat exchanger fouling in the United Kingdom is at least 500 million pounds. Fouling related problems in U.S. refinery heat exchangers cost at least 1.4 billion dollars annually; in the entire non-communist world, the cost is at least 4.4 billion dollars annually [11]. Of course, biofouling is only a part, albeit a significant part, of the overall problem.

There are very convincing data from laboratory and field experiments that show the industrial significance of biofouling on energy loss economics. Heat transfer resistance is considered to be the most significant manifestation of energy losses but the economic impact of fluid frictional resistance due to biofilm formation is often underestimated.

Biofilms can cause dramatic frictional drag on

moving fluids because the deposits significantly increase surface roughness [24]. For example, Knudsen [14] cites an interesting study of biofouling in Ocean Thermal Energy Conversion (OTEC) heat exchangers. The biofilm deposit was only 25 μm thick yet it reduced the capacity of OTEC condensers and evaporators by 32%. Picologlou [24] cites several examples including an interesting case of biofouling in a water-supply pipeline. The original capacity of an 80 km long, 600 mm I.D. pipeline was reduced 55% because of a 650 μm thick biofilm.

Biofilm Monitoring

The adherent bacteria present in a biofilm are the primary cause of biofouling problems [9-13,16,18,19 21,27-29]. Unfortunately, biofilm samples are difficult to obtain, particularly in heat exchanger systems. The earliest microscopic observations of biofilms from tubular flow systems were of pipe sections or scraped biofilm specimens. The biofilms were presumably distorted in the former and disoriented in the latter so that only limited conclusions were obtained. Microscopes have been mounted directly onto glass tubular flow systems in order to visualize biofilm development by bright-field microscopy [3,4]. This approach was successful in spite of the inherent problem of resolution.

The practical consequence of these limitations has forced systems managers to rely heavily on indirect indicators of surface biofouling. For example, the planktonic (free-floating) bacterial density in the bulk water is frequently used as a measure of biofilm accumulation on surfaces in contact with the water. More recently, the use of replaceable sampling surfaces has allowed more detailed observation of relatively undisturbed portions of fouling biofilms using either light or scanning electron microscopy. These sampling systems made it possible to correlate biofilm development with the physical manifestations of fouling. Details of these sampling systems are presented in Chapter 10.

Effects of Fluid Frictional Resistance

Fluid frictional resistance monitoring is an indirect approach that has been used to evaluate biofilms in some industrial water systems [8,16,19, 20,24,28]. The fluid frictional resistance encountered by a moving fluid in a tube can be monitored by calculating a directly proportional, dimensionless friction factor f [26]:

$$f = \frac{2\ \Delta PD}{v^2 pL}$$

where ΔP is the pressure drop across tube length L with diameter D, v is the average linear velocity of the fluid and p is the density of the fluid. Picologou *et al*. [24] showed that increases in fluid frictional resistance of bacterial biofilms were due to increased surface roughness such that a change in f could be measured when the biofilm extended beyond the viscous sublayer. On the basis of their observations, it was determined that a significant portion of the frictional resistance was due to the dynamic motion of microbial filaments extending into the bulk fluid. Others have shown that filamentous growth of bacteria in biofilms contributes significantly to increased fluid frictional resistance [20].

Figure 1 schematically demonstrates the effect of fouling biofilm formation on the fluid frictional resistance monitor, f. Biofilms formed in a laboratory scale heat exchanger operated at low flow (4.5 ft/s, Reynolds number = 25,000) and at high flow (8.7

ft/s, Reynolds number = 47,000) were shown to cause an increase in f (Figure 2). Laboratory experiments during which biofilm development has been followed with both microbiological analyses and with fluid frictional resistance monitoring have contributed significantly to our understanding of biofilm development (see Chapter 2).

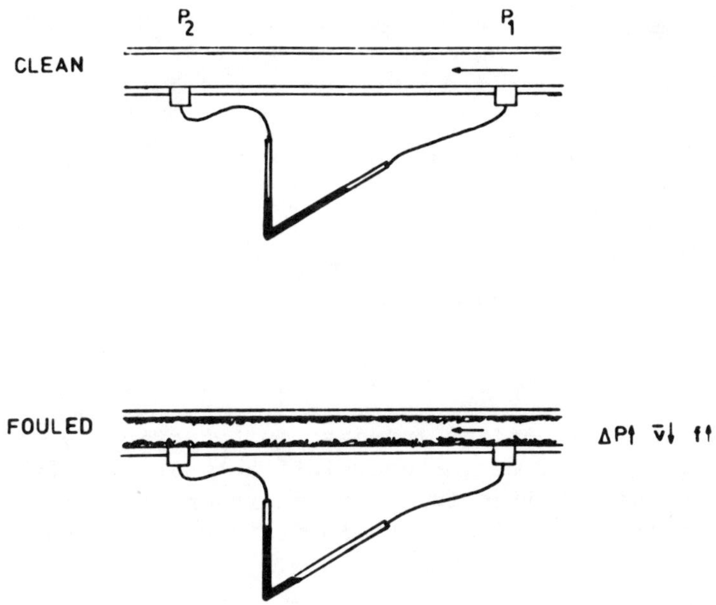

FIGURE 1. Fouling biofilm development (schematic). Two pressure ports (P_1 and P_2) were used to determine the pressure drop (ΔP) caused by biofilm formation. The average fluid velocity (v) was determined using volumetric flow meters. The friction factor, f, increased as a result of biofilm development.

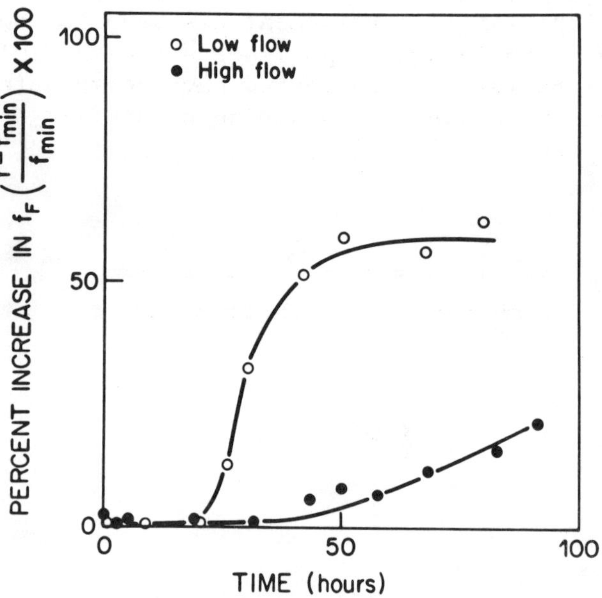

FIGURE 2. The percent increase in f during low and high velocity mixed culture biofilm runs. Biofilm development resulted in a more rapid onset of fouling (shorter lag phase) during low fluid velocity runs.

The calculation of the fluid frictional resistance, f, requires that ΔP and v be obtained from experimental results of measured values for cm of Hg and volumetric flow rate. Data from a laboratory biofilm experiment will serve to illustrate the calculation:

ΔP = cm Hg x v or 13.6 g/cm^3 x 980.7 cm/s^2

D = 1.422 cm

A = 1.589 cm^2

L = 174.7 cm

p = 0.9959 g/cm^3 (at 29^oC)

v = Q/A, where Q is volumetric flow rate (cm^3/s) and A is cross sectional area (cm^2)

$$Q/A = \frac{Q \times 10^6 \ cm^3}{1.589 \ cm^2 \times min}$$

$$v^2 = Q^2 \times (1 \times 10^8) \ cm^2/min^2$$

$$f = \frac{\Delta P}{Q^2} \times \frac{2 \times 13.6 \times 980.7 \times 1.422}{174.7 \times 0.9959} \times 10^{-8}$$

$$f = \frac{\Delta P}{Q^2} \times 2.18 \times 10^{-6} \quad \text{(a dimensionless factor)}$$

The sensitivity of f is usually limited by the measurement of ΔP. An inclined mercury manometer can usually detect a change of about 0.05 cm Hg. Thus, the smallest change measurable on the inclined mercury manometer (angle = 45°) can be calculated as:

$$0.05 \text{ cm Hg} + 0.05\ (0.707) \text{ cm Hg} = 0.085 \text{ cm Hg}$$

where the value 0.077 refers to sin 45°. Therefore, the smallest change in f measurable at the low velocity is about:

$$\frac{0.085 \text{ cm Hg} \times 2.18 \times 10^{-6}}{(0.0131 \text{ meters}^3/\text{min})^2} = 1.08 \times 10^{-3} \text{ units}$$

The smallest change in f measurable at the highest velocity is about:

$$\frac{0.085 \text{ cm Hg} \times 2.18 \times 10^{-6}}{(0.0252 \text{ meters}^3/\text{min})^2} = 2.92 \times 10^{-4} \text{ units}$$

The most profound effect of biofilm formation in industrial water cooling systems is on heat transfer. Very thin (microscopic) biological deposits on the walls of a heat exchanger can result in significant reductions in cooling efficiency [15,28]. The heat flux across a solid surface is described by Fourier's law:

$$q/A = -k \frac{dT}{dy} = U(\Delta T)$$

where q/A is the heat flux, k is the thermal conductivity of the solid, dT/dy is the temperature gradient across the solid with respect to a wall of thickness, y, and U is the overall heat transfer coefficient.

Heat transfer monitors can be used to follow the development of fouling biofilms in the laboratory and in the field [15,17,23]. A common approach to monitor the efficiency of industrial water cooling systems is by the "cleanliness factor" [1,23,25]:

$$CF = U_t/U_o$$

where CF equals the ratio of the overall heat transfer coefficient at any time, t, and at t = 0 (design specifications). The overall heat transfer coefficient is:

$$U = \frac{q/A}{(T_w - T_b)} = \frac{1}{R_w + R_f}$$

where T_w is the heat exchanger wall temperature (average), T_b is the bulk fluid temperature, R_w and R_f are the resistances to heat transfer due to the wall and the fouling deposits on the wall [15]. The resistance due to fouling, R_f, can be correlated with biofilm development as is demonstrated by the data given in Figure 3. It is evident from the examples given above that correlations between quantitative surface microbiology and energy-related parameters can be obtained.

Field devices for fluid frictional resistance and heat transfer resistance monitoring have become more generally available and useful in recent years. The Cooling Tower Institute (P.O. Box 73383, Houston, TX, 77273) has recently published guidelines for the selection and use of on-line fouling monitors along with a review of commercially available instruments [22]. The Electric Power Research Institute (EPRI, P.O. Box 10412, Palo Alto, CA, 94303) has produced a report describing the status of biofouling detection monitoring devices with recommendations for performing field evaluations [2].

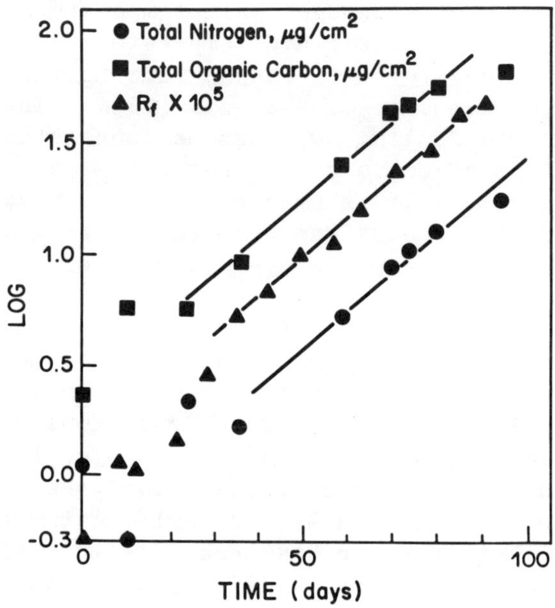

FIGURE 3. The exponential increase in heat transfer resistance, total organic carbon and total nitrogen with time on titanium heat exchanger surfaces. The experiment was done in ocean-water on an Ocean Thermal Energy Conversion heat exchanger model. Note that all three parameters have similar slopes.

References

1. Betz handbook of industrial water conditioning. 1980. 8th ed., Betz Laboratories Inc., Trevose, PA.

2. Biofouling detection monitoring devices: status assessment. 1985. Electric Power Research Institute CS-3914, Research Reports Center, Palo Alto, CA.

3. Bryers, J.D., and W.G. Characklis. 1981. Early fouling biofilm formation in a tubular flow system: overall kinetics. Water Res. 15:483-491.

4. Bryers, J.D., and W.G. Characklis. 1982. Processes governing primary biofilm formation. Biotech. Bioeng. 24:2451-2476.

5. Characklis, W.G. 1973. Attached microbial growths-I. Attachment and growth. Water Res. 7:1113-1127.

6. Characklis, W.G. 1981. Fouling biofilm development: a process analysis. Biotechnol. Bioengin. 23:1923-1960.

7. Characklis, W.G., and K.E. Cooksey. 1983. Biofilms and microbial fouling. Adv. Appl. Microbiol. 29:93-138.

8. Colturi, T.F., and K.J. Kozelski. 1983. Corrosion and biofouling control in a cooling tower system with demineralized water make-up. Paper no. 280, National Association of Corrosion Engineers annual meeting, NACE Publications, Houston, TX

9. Crandall, H.E. 1986. Microbiological test methods in association with cooling towers. Paper no. TP-86-7, Cooling Tower Institute. Houston, TX.

10. Ferguson, R.J. 1981. Determination of seasonal variations in fouling factors and apparent slime thickness. Paper no. TP-237A, Cooling Water Institute, Houston, TX.

11. Helton, D.C., D.A. Johnson, and G.W. Hanks. 1985. New approaches to monitoring the performance and control of cooling water programs. Paper no. TP-85-12, Cooling Tower Institute, Houston, TX.

12. Johnson, C., and M. Howells. 1981. Biofouling: new insights, new weapon. Power 125(4):90-91.

13. Kajdasz, R., R.V. Einstman, and L. Young-Bandala. 1984. Biocide efficacy with respect to sessile and planktonic organisms. p. 85-93. In Proceedings of the international water conference. IWC-84-60., Pittsburgh, PA.

14. Knudsen, J.G. 1981. Fouling of heat transfer surfaces. p. 373-425. In P.J. Marto and R.H. Nunn (ed.), Power condenser heat transfer technology, Hemisphere Publishing Co., New York.

15. Knudsen, J.G. 1981. Apparatus and techniques for measurement of fouling of heat transfer surfaces. p. 57-82. In E.F.C.Somerscales and J.G. Knudsen (ed.), Fouling of heat transfer equipment, Hemisphere Publishing Co., New York.

16. Kozelski, K.J. 1983. Field experience with a simple cooling water biofilm monitoring device. p. 447-454. In Proceedings of the international water conference, IWC-83-46, Pittsburgh, PA.

17. Liebert, B.E., L.R. Berger, H.J. White, W.F. McCoy, J.A. Berger, and J.Larsen-Basse. 1980. The effects of biofouling and corrosion on heat

transfer measurements. p. 12.71-12-710. In G.L. Dugger (ed.), Proceedings of the 6th OTEC conference, Systems Branch, Division of Central Solar Technology, U.S. Dept. of Energy, Washington, D.C.

18. McCoy, J.W. 1980. Microbiology of cooling water. Chemical Publishing Co., New York.

19. McCoy, W.F., and J.W. Costerton. 1982. Fouling biofilm development in tubular flow systems. Dev. Ind. Microbiol. 23:551-558.

20. McCoy, W.F., and J.W. Costerton. 1982. Growth of sessile Spaerotilus natans in a tubular recycle system. Appl. Environ. Microbiol. 43:1490-1494.

21. McCoy, W.F., J.E. Ridge, and E.S. Lashen. 1986. Efficacy of biocides in laboratory model cooling towers. Materials Performance 25:9-14.

22. Moore, W.E., and G.F. Hays. 1985. Guidelines for selection and use of on-line fouling monitors. Paper no. TP-85-2, Cooling Tower Institute. Houston, TX.

23. Nalco Chemical Co. 1979. The Nalco water handbook. McGraw-Hill Book Co., New York.

24. Picologou, B.F., N. Zelver, and W.G. Characklis. 1980. Biofilm growth and hydraulic performance. J. Hydraul. Div., Proceedings of the American Society of Civil Engineers 106:733-746.

25. Principles of industrial water treatment. 1978. Drew Chemical Corp., Boonton, NJ.

26. Schlichting, H. 1968. Boundary layer theory. McGraw-Hill Book Co., New York.

27. Song, P., and M.G. Trulear. 1986. Identification of cooling tower wood attack and methods of control. Paper no. TP-86-10, Cooling Tower Institute, Houston, TX.

28. Strauss, S.D., and P.R. Puckorius. 1984. Cooling-water treatment for control of scaling, fouling, corrosion. Power 128(6):S1-24.

29. Trulear, M.G., and R.L. Wetegrove. 1984. Biofilm dynamics. Paper no. 99, Proceedings of the National Association of Corrosion Engineers annual meeting, NACE Publications, Houston, TX.

CHAPTER 9. STRATEGIES FOR THE TREATMENT OF BIOLOGICAL FOULING

W.F. McCoy, Great Lakes Chemical Corporation,
West Lafayette, IN

Effective control of biofouling generally requires adequate diagnosis of the problem followed by application of an appropriate treatment technique. In this chapter, the removal of biofilms from industrial water systems will be discussed in terms of physical and chemical methods that are currently in practice. Since considerable information exists on these topics, the reader is directed to the references for those details which are not discussed below.

Systematic analytical microbiology procedures such as those described in Chapter 14 and fluid frictional resistance monitoring techniques such as those described in Chapter 8 are essential for the assessment and control of biofilm fouling. Fluid frictional resistance monitoring techniques have been successfully used to assess the efficacy of biocide treatments under actual operating conditions in the industrial environment. Kozelski [22] and Colturi and Kozelski [8] have presented evidence from the field that shows on-line fluid frictional resistance monitoring that was sensitive enough to assess the efficacy of chemical biofouling programs.

Effects of Surface Characteristics

Although bacterial biofilms will develop on virtually any solid surface exposed to a contaminated aqueous environment, some surfaces discourage bacterial colonization better than others. Figure 1 shows the characteristics of biofilms that developed on tubing of four metallurgically distinct heat

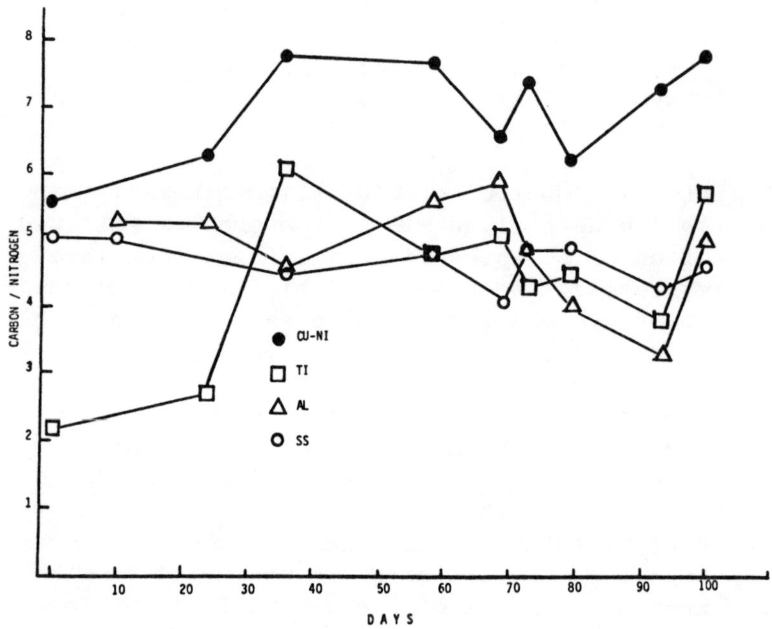

FIGURE 1. Ratio of carbon to nitrogen in the fouling layer on four metallurgically distinct heat exchanger surfaces.

exchangers. The copper/nickel alloy (90:10) was fouled by a deposit that was significantly richer in carbon (presumably of biological origin). Microscopic and biochemical evidence from this and other studies show that carbon-rich extracellular polysaccharides may afford the cells some protection from the toxic effects of the metal surface [10,28].

Coatings applied to surfaces may inhibit biofilm formation to some extent by inclusion of chemical biocides in their formulation. Biocides used for this purpose include organo-metallic, organo-sulfur, phenolic, and amine-based compounds. However, there are no practical examples of industrial biofouling control achieved solely by the application of surface coatings. Nevertheless, somewhat limited control of biofouling on certain surfaces is feasible with coatings technology (e.g., painting the hulls of ships). Recently, TAFA Inc. (Concord, MA) described a thermal atomization method of copper deposition for coating concrete surfaces in sewage clarifying tanks that prevents the growth of algae.

Shear

There are several practical, physical techniques for removing biofilms from surfaces that rely on the application of shearing forces. Table 1 lists the most frequently applied practical mechanical techniques for removing fouling deposits from surfaces.

Figure 2 illustrates the effect of brushing on heat transfer resistance in fouled heat exchanger tubes. A very common observation is that the rate of "regrowth" is often significantly increased following application of physical treatments (note the rate of increase in R_f and protein levels on stainless steel heat exchangers before and after day 60 in Figure 2). In another field trial evaluation, Berger and Berger [6] presented microscopic evidence to explain certain regrowth phenomena following passage of rubber sponge balls through fouled heat exchangers. The balls "combed" the biofilm and reduced the diversity of microorganism and filamentous appearance. The sponge ball treatment did not achieve complete removal of the biofilm. An important consideration with respect to physical biofilm removal techniques is the possible acceleration of corrosion rates due to the abrasive removal of protective oxide film [3]. Costerton [11] recently described a novel approach to remove biofilms from heat exchanger surfaces using fluid shear coupled with ice nucleation. It remains to be shown whether this technique is economically feasible, however.

TABLE 1. Common physical methods to clean biofouling deposits from surfaces.

Technique	Comments
Flushing	simplest method; very limited efficacy
Backwashing	effective for loosely adherent films in tubes, on filters, and in ion-exchange resin beds
Air Bumping	very limited efficacy
Non-abrasive and/or Abrasive Sponge Balls	extensively used in industry; demonstrated efficacy
Sand Scouring	difficult to control abrasive effects on metal surfaces
Brushing	expensive; very effective

Filtration

Side-stream filtration of industrial process water through sand or anthracite is sometimes used to reduce the amount of suspended dirt and debris;

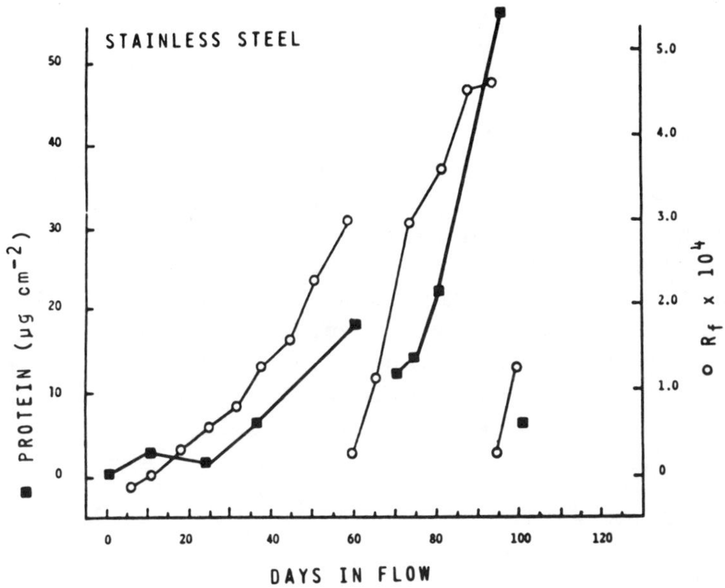

FIGURE 2. Protein in the fouling layer and heat transfer resistance of tubing as a function of time of exposure to flowing sea water at 6.6 ft/s. R_f measurements are given in conventional engineering units ($H^oF\ ft^2\ BTU^{-1}$), where H refers to heat transfer coefficient. To convert to international units ($H^oC\ M^2\ Joule^{-1}$) multiply by 5.66×10^{-3}. The tubes were cleaned by brushing at hour 60 and at hour 95.

however this process is ineffective at limiting biofouling activities on surfaces in industrial systems. Membrane filtration, and more recently, reverse osmosis membrane filtration, have been used to purify tertiary-treated sewage and other heavily contaminated waters (Chapter 6). Membrane filtration is an essential process in ultrapure water technology.

Radiation

There is a current debate on the merit of radiation treatments for the preservation of foods and beverages. The notion that such methods could be applied to industrial process waters for the purpose of limiting biofilm formation does not appear at all practical. Another form of radiation, ultraviolet light, is quite practical and often used in industrial water systems. Biofouling control by application of UV radiation is discussed in Chapter 5.

Chemical Treatments

Traditionally, chemical compounds used for biofouling control in industry have been categorized as either oxidizing or non-oxidizing biocides. With the exception of hydrogen peroxide and ozone, all of the oxidizing agents used as biocides contain halogens.

Oxidizers (Halogens)

Chlorine is the most frequently applied industrial water system biocide. Chlorine is usually administered to water as a gas:

$$Cl_2 + H_2O \rightarrow HOCl + H^+ + Cl^-$$

where the K_{eq} at 25°C is about 2.5×10^{-3}. Most industrial water is sufficiently buffered such that chlorination does not appreciably lower the pH. Hypochlorous acid, HOCl, is an effective biocide at the low use-levels required for most industrial applications. The acid dissociates to hypochlorite ion, which is significantly less biocidal [16,25,40, 41]:

$$HOCl \rightarrow H^+ + OCl^-$$

where the pK_a at 25°C is 7.4. It is generally believed that low levels of free available residual chlorine ($HOCl + OCl^-$) used for industrial applications are much less effective in alkaline waters compared to neutral or acid waters because of this acid dissociation reaction.

Combined available residual chlorine refers to that which is nitrogen-bound; i.e., chloramines [17, 18]:

$$R_2N\text{-}Cl$$

where R can be H, Cl, or any organic moeity. Chloramines are generated in most industrial chlorination processes but they are significantly less effective than the free residual [29,42] and are much more volatile. For example, Holzworth _et al_. [19,20] presented evidence that the chloramines are probably removed from water-recirculation systems by flashoff so quickly that they are essentially unavailable as disinfectants. The process to complete all of the chloramination reactions is known as breakpoint chlorination [30]. Chlorination in industrial water treatment is usually applied either as an extremely low continuous dose (0.2-0.5 ppm total available residual) or as a shock-dose (1-2 ppm free residual) at regular intervals. Shock doses of chlorine (50-100 ppm) are commonly used in the treatment of potable water distribution and storage facilities. Such high levels of chlorine must be inactivated by sodium thiosulfate or other appropriate agents prior to disposal.

In certain chlorination applications, organic compounds that deliver oxidizing halogen are used instead of gaseous chlorine because of safety and convenience. Organic halogen-delivery is more expensive than gas chlorination. The major group of organic chlorine donors is the chloroisocyanurates; there are at least five commercially significant products but the two most important are trichloro-s-triazinetrione and sodium dichloro-s-triazinetrione (Figure 3). Chloroisocyanurates are white, crystalline solids that have a distinct chlorine odor. These compounds find extensive application as swimming pool disinfectants and in industrial cleaning products.

Chlorine combines readily with the nitrogen in cellular proteins. Gross changes due to chloramination of cell proteins render chlorine toxic to all living systems. Loss of membrane integrity leading to leakage of intracellular potassium ions [15] and disruption of membrane-mediated cell processes (e.g., active transport) are probably significant inactivation mechanisms. Oxidation of essential reducing enzymes in respiration, primarily at sulfhydryl groups, is also known to occur, but this is probably not the primary cause of cell death.

The salts of hypochlorous acid have found application as biocides in certain water treatment applications. Sodium hypochlorite, familiar as the active ingredient in household bleach, is predominant among these but its application is limited by the poor stability of concentrated solutions. Solid forms of calcium hypochlorite ("dry chlorine") are stable and used extensively as industrial sanitizers and swimming pool disinfectants. Of course, the addition of large amounts of calcium prohibits the use of dry chlorine in many applications.

Another biofouling control method based on

chlorine chemistry utilizes the water-soluble gas, chlorine dioxide, (ClO_2). Unlike chlorine gas, ClO_2 does not react with water. The biocide is usually generated on-site from sodium chlorite:

$$Cl_2 + H_2O \rightarrow HOCl + HCl$$

$$HOCl + HCl + 2NaClO_2 \rightarrow 2ClO_2 + 2NaCl + H_2O$$

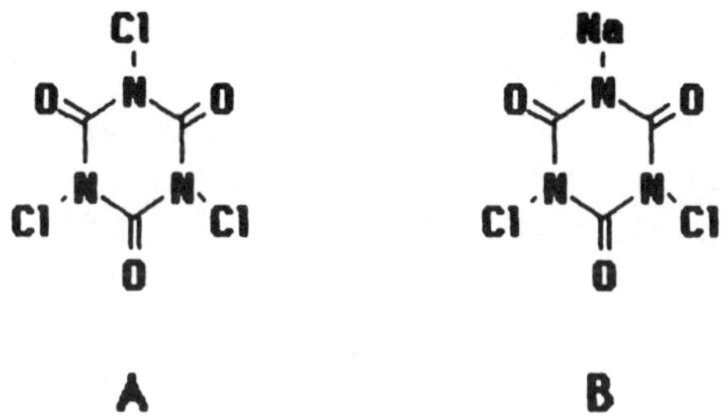

FIGURE 3. The two most commercially important chloroisocyanurates are (A) trichloro-s-triazinetrione and (B), sodium dichloro-s-triazinetrione.

Thus, the bioactive chemical is not the oxyacid (HOCl) but rather, a gaseous oxide of chlorine. The sodium chlorite reaction is generally run in a side-stream reactor downstream from the chlorinator [37]. Chlorine dioxide can be generated from sodium chlorite without chlorine gas by the addition of acid [2,37]:

$$5H^+ \text{ (acid)} + 5NaClO_2 \rightarrow 4ClO_2 + HCl + 2H_2O + 5Na^+$$

or by the addition of bleach [37]:

$$2NaClO_2 + NaOCl + H_2O \rightarrow 2ClO_2 + 2NaOH + NaCl$$

Compared to hypochlorous acid, the efficacy of chlorine dioxide in water is less affected by pH, ammonia concentration, and hydrocarbon contamination. The mode of action of chlorine dioxide is not clearly known, but recent studies indicate that the primary lethal event may be due to the loss of cell-membrane permeability control which leads to disruption of the trans-membrane ionic gradient [5]. ClO_2 production can be hazardous: the gas is explosive under certain conditions. Appropriate safety practices should be followed in the production and induction of chlorine dioxide.

Bromination is an important alternative to chlorination in the water treatment industry because of the unique chemistry of bromine [31]. The element is a fuming red liquid at standard temperature and pressure; it is soluble in water (about 33.6 g/l at 25°C) and reacts to form the oxyacid, analogous to chlorine:

$$Br_2 + H_2O \rightarrow HOBr + H^+ + Br^-$$

where the K_{eq} at 25°C is about 1.6×10^{-8}. The acid dissociation is analogous to hypochlorous acid:

$$HOBr \rightarrow H^+ + OBr^-$$

where the pK_a at 25°C is about 8.8.

The acid dissociation constant for hypobromous acid is more than one pK_a unit higher than the constant for hypochlorous acid (8.8 and 7.4 at 25°C, respectively). The practical consequence of this fact is that hypobromous acid will be significantly less dissociated compared to hypochlorous acid in water with pH between 7 and 10. The low use-levels of oxidizing halogen are most effective as disinfectants in the undissociated hypohalous form; therefore, bromine will perform significantly better than chlorine in alkaline water. This principle has been adequately demonstrated in practice [4,8,9,14,21,22, 24,31].

In addition to the acid dissociation phenomenon, bromine chemistry offers a considerable advantage over chlorine with respect to the formation of haloamines. The efficacy of the bromamines is significantly greater than the chloramines [31] and they are less persistant in the environment. Brominated carbon compounds, most notably bromoform, are known to be less persistant than the analogous chlorine compounds and thus, the environmental problem of trihalogenated methane (THM) contamination is believed to be considerably less significant in bromination processes [7,23].

Although liquid elemental bromine has been aspirated directly into water systems to control biofouling [1], the method is rarely employed because of the limited solubility of bromine as well as safety considerations. Currently, there are three practical methods to administer oxidizing bromine to industrial waters: 1) N-halo organic donors, 2) activated sodium bromide, and 3) bromine chloride.

The most successful organic bromine-releasing agent for industrial water treatment is the N-halo compound, 1-bromo-3-chloro-5,5-dimethylhydantoin (Figure 4). The compound is a slightly water-soluble solid (about 1.5 g/l at 25°C) that can be rather easily applied to industrial water by means of a simple side-stream reactor [24,31]. Hypohalous acid is generated when the solid dissolves in water:

$$C_5H_6BrClN_2O_2 + 2H_2O \rightarrow C_5H_8N_2O_2 + HOBr + HOCl$$

A convenient method to generate HOBr is to "activate" sodium bromide by oxidation. This technique is most appropriate in systems that employ gas chlorine or hypochlorite processes [32]. A solution of the salt, NaBr, is added downstream from

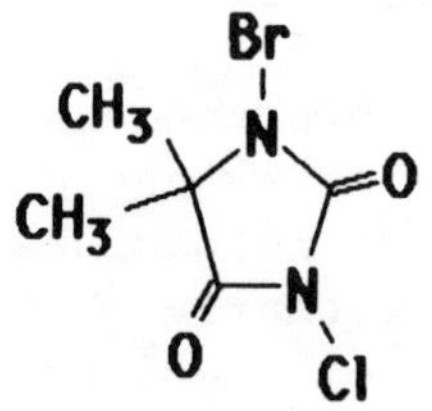

Figure 4. 1-Bromo-3-chloro-5,5-dimethylhydantoin generates HOBr and HOCl upon dissolution in water.

the chlorinator such that the following oxidation-reduction reactions occur:

$$HOCl + Br^- \rightarrow HOBr + Cl^-$$

$$OCl^- + Br^- \rightarrow OBr^- + Cl^-$$

$$R_2NCl + Br^- \rightarrow R_2NBr + Cl^-$$

This process is known to considerably enhance the efficacy of chlorination processes and, in practice, it has allowed certain large-scale chlorine users to satisfy the governmental requirements for extremely low-level release of oxidizing halogen into the environment [9,31,34].

The most economical way to deliver hypobromous acid to industrial water is with bromine chloride [14, 31],

$$BrCl + H_2O \rightarrow HOBr + H^+ + Cl^-$$

Bromine chloride is an interhalogen equilibrium product that results from the equimolar reaction of elemental bromine and chlorine. The product is a liquid at ambient temperatures only under considerable pressure (20 psia) and therefore, requires special handling and feeding equipment. The benefits of using BrCl rather than its respective elements are: a) much lower corrosivity, b) much higher solubility in water, and c) the least expensive way to generate hypobromous acid in water.

Non-oxidizers

This category refers to chemical biocides that are relatively non-reactive and thus, more compatible with strong reducing agents in water treatment applications. For most industrial applications, a non-oxidizing biocide is any commercial product that does not generate oxidizing-halogen or any other strong oxidant (e.g., hydrogen peroxide or ozone) in water. Such generalizations tend to grossly over simplify the facts; many commercial "non-oxidizers" (e.g., DBNTA) are indeed, rather good oxidizing agents. The distinctions are not clear-cut, so that for the purpose of the discussion presented here, the major "non-oxidizing" biocides used in the water treatment industry will be considered separate from the oxidizing biocides described above.

Table 2 lists the major groups of non-oxidizing biocides with a few of the most important products in each group [12,13,25-27,33,35,36,38,39]. Common abbreviations for these molecules are included in this table and will be used throughout the text. This list of biocides does not include every commercial product currently available; omission from the list does not imply inferior performance in any way.

Non-oxidizing biocides are usually used to enhance the oxidizing biocide program. The former are desirable where oxidizing biocides occasionally cannot be used in certain water treatment programs because of chemical incompatibility or because of high halogen demand. In most cases, the non-oxidizer is used to treat specific microbiological problems that

TABLE 2. "Non-oxidizing" biocides used for industrial water treatment.

Organobromine
- 2-2-Dibromo-3-nitropropionamide (DBNPA)
- β-Bromo-β-nitrostyrene (BNS)
- 2-Bromo-2-nitropropane-1,3-diol (BNPD)

Organometallics
- Tributyltin oxide (TBTO)

Organonitrogen
- Alkyldimethylbenzylammonium chloride (ADBAC Quat)
- Dialkyldimethylammonium chloride
- N-coco-trimethylenediamine
- Dodecylguanidine hydrochloride

Organonitrogen-sulfur
- 5-Chloro-2-methyl-4-isothiazolin-3-one/ 2-methyl-4-isothiazolin-3-one
- Tetrahydro-3,5-dimethyl-2H,1,3,5-thiadiazine-2-thione (DMTT)
- Disodium ethylene bisdithiocarbamate

Organosulfur
- Methylene bisthiocyanate (MBT)
- Bis (trichloromethyl) sulfone
- Hexachlorodimethylsulfone

Miscellaneous aldehydes
- Acrolein
- Formaldehyde
- Glutaraldehyde

occur in spite of regular oxidizing biocide treatments. An example of such a problem is the extremely heavy algal biofilms on cooling tower distribution boxes, that frequently develop during the summer months. Most often, a non-oxidizer will be administered in slug-doses at regular intervals (e.g., two or three times per week) in an attempt to control the development of biofilms.

Each of the products in Table 2 have specific use-levels, application protocols, and safe-handling procedures that are not considered here. For more information on these subjects, the reader is refered to the water treatment companies that act as distributors for the manufacturers of these products. In many cases, end-users benefit by employing the expertise of a water treatment specialist in the areas of worker and environmental saftey and biocide efficacy.

Non-oxidizing biocides are often used in combination with chemical surfactant/dispersant systems. In fact, some biocides (e.g., MBT and TBTO) are really not practical for water treatment applications without a dispersant (to insure an effective concentration throughout the system) and a surfactant (to encourage penetration of the biocide into the biofilm). Proprietary dispersant/surfactant systems used to formulate non-oxidizing biocides for industrial applications are another important area of expertise that water treatment specialists can bring to the end-user.

There are many examples of synergistic combinations of non-oxidizing biocides. Important examples include MBT plus sulfones and MBT plus dodecylguanidine hydrochloride. Studies of synergistic interactions between non-oxidizing biocides are an important and fruitful area of industrial microbiological research.

An informative way to categorize the non-oxidizing biocides is by rough approximation of their modes of action. It must be emphasized that the precise mechanism of efficacy is usually not known.

Nevertheless, one can roughly categorize the non-oxidizers based on the following target sites. Enzyme poisons (e.g., MBT, TBTO, DBNPA, carbamates, BNS) inhibit protein-catalyzed biochemical reactions such as respiration and active transport. Cell wall or cell membrane disruptors (e.g., ADBAC quats, dodecylquanidine hydrochloride, alkyldiamines) affect the integrity of the semi-permeable membrane. General protein inactivators (e.g., acrolein, formaldehyde, glutaraldehyde) react in a variety of ways to inhibit enzymes, cross-link macromolecules in membranes, or cause cell agglutination.

In certain cases, knowledge of the general mode of antimicrobial action for industrial products has been useful in designing biocide treatment programs and in interpretation of field results. A good example of benefit derived from this kind of information is shown by example as follows. Biocide "A" is a broad-spectrum (i.e., effective against many different types of microorganisms) non-oxidizer that inhibits active transport and macromolecular syntheses by interaction with cellular proteins. Biocide "B" is also a broad-spectrum, non-oxidizer, but it effectively "poisons" specific oxidases (enzymes) in the respiratory system of microorganisms. Standard bottle tests for efficacy from the field (e.g., ASTM E645) will probably show that the required contact time for "A" will be relatively long (at least 6 hours) compared to "B" (probably less than 1 or 2 hours) when the industrial use-level of active ingredients is administered. This prediction is based on knowledge of the modes of action: "A" performs as a slow-acting, perhaps even bacteriostatic biocide while "B" actually suffocates microbial cells by inhibiting respiration. Biocide "A" must be administered such that its effective level is maintained for an extended period of time. Biocide "B" should be frequently shock-dosed to achieve a quick-kill; maintenance of an effective level for more than a few hours would not necessarily be required. In practice then, either Biocide "A" or "B" may be used successfully by exploiting knowledge of their respective inhibitory action.

References

1. Albright, J.C. 1948. Liquid bromine removes obstinate algae from 10,000 gpm tower for $2.10/day. Petrol. Prog. 3:421-422.

2. Alliger, H. 1984. Germ-killing composition and method. U.S. Patent Re. 31,779.

3. Bartz, J.A., R.J. Bell, Y.G. Mussalli, D.H. Pope, and M.D. Rosen. 1987. Results of steam condenser fouling and corrosion studies. Paper presented at the 1987 American Society of Mechanical Engineering annual winter meeting. Correspondance: J.A. Bartz, Electric Power Research Institute, P.O. Box 10412, Palo Alto, CA, 94303.

4. Beckwith, T.D., and J.R. Moser. 1933. Germicidal effectiveness of chlorine, bromine, and iodine. J. Am. Water Works Assoc. 25:367-374.

5. Berg, J.D., P.V. Roberts, and A. Matin. 1986. Effect of chlorine dioxide on selected membrane functions of *Escherichia coli*. J. Appl. Bacteriol. 60:213-220.

6. Berger, L.R., and J.A. Berger. 1986. Countermeasures to microbiofouling in simulated ocean thermal energy conversion heat exchangers with surface and deep ocean waters in Hawaii. Appl. Environ. Microbiol. 51:1186-1198.

7. Bouwer, E.J., B.E. Rittmann, and P.L. McCarty

1981. Anaerobic degradation of halogenated 1- and 2-carbon organic compounds. Environ. Sci. Tech. 15:596-599.

8. Colturi, T.F., and K.J. Kozelski. 1983. Corrosion and biofouling control in a cooling tower system with demineralized water make-up. Paper no. 280, National Association of Corrosion Engineers annual meeting, NACE Publications, Houston, TX.

9. Conley, J.C., and E.H. Puzig. 1987. Bromine chemistry, an alternative to dechlorination. Proceedings of the condenser technology symposium of the Electric Power Research Institute. In Press.

10. Corpe, W.A. 1975. Metal-binding properties of surface materials from marine bacteria. Dev. Ind. Microbiol. 16:249-255.

11. Costerton, J.W. 1983. Biofilm removal. U.S. Patent No. 4,419,248.

12. Friend, P.L., and W.K. Whitekettle. 1980. Biocides and water cooling towers. Dev. Ind. Microbiol. 21:123-131.

13. Gorman, S.P., and E. Scott. 1980. A Review: Antimicrobial activity, uses and mechanism of action of glutaraldehyde. J. Appl. Bacteriol. 48:161-190.

14. Groninger, G.D., and J.F. Mills. 1980. Disinfection of wastewater plant effluents using bromine chloride versus chlorine and its environmental impact. p. 733-740, In R.L. Jolley, (ed.), Water chlorination: environmental impact and health effects, Vol. 3, Ann Arbor Science,

Ann Arbor, MI.

15. Haas, C.N., and R.S. Englebrecht. 1980. Physiological alterations of vegetative microorganisms resulting from chlorination. J. Water Pollut. Cont. Fed. 52:1976-1989.

16. Hass, C.N., and S.B. Karra. 1984. Kinetics of microbial inactivation by chlorine: review of results in demand-free systems. Water Res. 18:1443-1449.

17. Hass, C.N. and S.B. Karra. 1984. Kinetics of microbial inactivation by chlorine: kinetics in the presence of chlorine demand. Water Res. 18:1451-1454.

18. Helton, D.C., D.A. Johnson, and G.W. Hanks. 1985. New approaches to monitoring the performance and control of cooling water programs. Paper no. TP-85-12, Cooling Tower Institute, Houston, TX.

19. Holzworth, G., R.G. Balmer, and L. Soni. 1984. The fate of chlorine and chloramines in cooling towers: Henry's Law Constants for flashoff. Water Res. 18:1421-1427.

20. Holzworth G., R.G. Balmer, and L. Soni. 1984. The fate of chlorine in recirculating cooling towers: field results. Water Res. 18:1429-1435.

21. Johnson, C., and M. Howells. 1981. Biofouling: new insights, new weapon. Power 125(4):90-91.

22. Kozelski, K.J. 1983. Field experience with a simple cooling water biofilm monitoring device. p. 447-454, _In_ Proceedings of the international water conference, IWC-83-46, Pittsburgh, PA.

23. Liden, L.H., D.K. Burton, L.H. Bongers, and A.F. Holland. 1980. Effects of chlorobrominated and chlorinated cooling waters on estuarine organisms. J. Water Poll. Cont. Fed. 52:173-182.

24. Macchiarolo, N.T., C.H. Johnson, and R.S. Hornack. 1980. A new biocide for cooling water systems. Paper no. TP-219-A, Cooling Tower Institute, Houston, TX.

25. McCoy, J.W. 1983. The chemical treatment of cooling water. 2nd ed., Chemical Publishing Co., New York.

26. McCoy, W.F., J.E. Ridge, and E.S. Lashen. 1986. Efficacy of biocides in laboratory model cooling towers. Materials Performance 25:9-14.

27. Meitz, A.K. 1984. Efficacy and decomposition of DBNPA in two cooling systems. p. 243-298, Proceedings of the international water conference, IWC-84-50, Pittsburgh, PA.

28. Mittelman, M.W., and G.G. Geesey. 1985. Copper-binding characteristics of exopolymers from a freshwater sediment bacterium. Appl. Environ. Microbiol. 49:846-851.

29. Montgomery, J.M. 1985. Water treatment principles and design. John Wiley & Sons, Inc., New York.

30. Palin, A.T. 1983. Chemistry and control of modern chlorination. LaMotte Chemical Products Company, Chestertown, MD.

31. Sergent, R.H. 1986. Enhanced water management using bromine chemistry. Paper no. TP-86-9,

Cooling Tower Institute, Houston, TX.

32. Shere, L., M.J. Kelley, and J.H. Richardson. 1962. Effect of bromide-hypochlorite bactericides on microorganisms. Appl. Microbiol. 10:538-541.

33. Sondossi, M., H.W. Rossmoore, and J.W. Wireman. 1986. The effect of fifteen biocides on formaldehyde-resistant strains of *Pseudomonas aeruginosa*. J. Ind. Microbiol. 1:87-96.

34. Strauss, S.D. 1985. Condenser biofouling control looms large in light of toxic-discharge deadline. Power 129(9):83-85.

35. Tanis, J.N. 1987. Procedures of industrial water treatment. Latan, Inc., Ridgefield, CT.

36. Tomlin, J.L. 1985. Hints on improving microbiological control in cooling water. Paper no. 295, Proceedings of the National Association of Corrosion Engineers annual meeting, NACE Publications, Houston, TX.

37. Ward, W.J. 1976. Chlorine dioxide: a new development in effective microbiological control. Paper no. TP-153A, Cooling Tower Institute, Houston, TX.

38. Wehner, D.C., and C.F. Hinz. 1971. Organic thiocyanates as industrial antimicrobial agents. Dev. Ind. Microbiol. 12:404-410.

39. Wetegrove, R.L., and F.C. Pocius. 1983. Controlling microorganisms in cooling-water systems. Chem. Engineer. 90(22):75-78.

40. Wheeler, W.W. 1978. A simplified application of

chlorine for biological control. Paper no. TP-187-A, Cooling Tower Institute, Houston, TX.

41. White, G.C. 1985. Handbook of chlorination. Van Nostrand Reinhold Co., New York, NY.

62. Wolfe, R.L., N.R. Ward, and B.H. Olsen. 1984. Inorganic chloramines as drinking water disinfectants: A review. J. Am. Water Works Assoc. 76(5):74-88.

PART II. LABORATORY METHODS

CHAPTER 10. SAMPLE COLLECTION AND PROCESSING

Planktonic Sampling

In order to obtain an accurate estimate of microbial numbers, the sample taken must be representative of the area being tested. Sampling errors account for a significant number of retests associated with microbial assays. Great care must be taken to ensure that the sample is free from extraneous contaminants. Additionally, microorganisms tend to multiply much more rapidly at increased temperatures. Since an estimate of microbial numbers present at the time of sampling is desired, the sample must be refrigerated immediately to prevent the occurrence of falsely high microbial counts. The sample valve should possess three attributes: a small oriface (<1/4"), a minimum residual volume of water (<2 ml), and a small internal surface area (e.g., a needle-type rather than a ball-type valve). Gasketed, globe-type valves should never be used for sampling.

Procedure

1. Clean the exterior surface of the sample port with a clean cloth soaked in 70% ethanol (or isopropanol).

2. Flush the sample port for three minutes with the sample spigot wide open. If the flow is reduced to avoid splashing, the sample port must be flushed at the lower flow rate for an additional three minutes.

3. Carefully remove the cap from a presterilized sample container, taking care not to touch the inside surfaces of the cap or the sample

container.

4. Slowly fill the container to the neck, taking care to avoid any splashing. A small void volume should be present at the top of the container to facilitate mixing. Carefully replace the cap.

5. If the inside of the container or cap is accidentally touched, or if any spashing occurs, take another sample.

6. After labeling the sample container with the sample location, time, date, and sampler's initials, immediately place it in an ice chest or refrigerated (ca. $4^{o}C$) container for transport.

7. Transport samples to the laboratory for processing within 6 h of sampling.

Biofilm Sampling

Coupons

Coupons are convenient to use for isolation and enumeration of bacteria that attach to surfaces. Coupons composed of material comparable to the surfaces in an industrial water system may be readily obtained from commercial suppliers or fabricated "in house". Depending on the extent of surface colonization by microorganisms, appropriate size coupons may be selected for optimal results. Evaluation of coupon fouling may be carried out by scanning electron microscopy, epifluorescence microscopy, or by cultural methods.

Sample Collection and Processing

Robbins Biofilm Sampler

On-line monitoring of process waters for biofouling activities can be accomplished using a biofilm sampler such as the Robbins Device (see schematic diagram on the inside back cover). Originally developed by J.W. Costerton of the University of Calgary in Alberta, this sampler is designed to provide replicate surfaces for studies of microbial attachment and subsequent biofilm formation. The sampler can be constructed of any machineable materials of construction that are compatible with the process waters under consideration. The removable sample studs should simulate the distribution piping materials of construction and piping surface finish characteristics.

Although it is not necessary to steam sterilize biofilm samplers prior to use, the units should be critically cleaned with a detergent followed by 70% ethanol and 0.2 μm-filtered water rinses. Prior to introduction of the sampler into the process stream, precleaned sample studs should be assayed for background (control) levels of the analytes of interest (viable and total bacteria, carbohydrate, protein, endotoxin, physical/chemical corrosion assays, etc). The lead-in section to the biofilm sampler should be designed such that turbulent flow is minimized, thus allowing even biofilm formation on all of the sample studs. In a typical experiment, the precleaned sampler is placed on-line for a period of 14-28 days. Sample studs are pulled at two-day intervals and assayed. In this way, qualitative and quantitative observations of biofilm formation may be made over the duration of the test period. If treatment efficacy data is desired, a number of samplers can be sidestreamed and challenged with chemical or physical treatments during the challenge period.

Anaerobic bacteria assays for sulfate-reducing bacteria should be conducted under a sterile nitrogen atmosphere utilizing the appropriate media (see also ch. 13). Biochemical assays may also be performed on biofilm scrapings. Typically, bacteria and biochemical results are normalized to a measure of surface area; e.g., μg protein/cm^2.

Biofilm Sample Processing

1. Replicate sample surfaces (SS) of appropriate composition are place in the system at various locations including areas where the water is fast-moving as well as stagnant.

2. After various periods of exposure to the water a SS is aseptically recovered from each test site and rinsed with sterile process water. Flamed forceps may be used to handle the coupon.

3. The biofilm is removed quantitatively from each SS with a sterile scalpel blade and transferred to a separate tube containing 10 ml of sterile process water or distilled water.

4. Each tube is placed in a water bath sonicator for 1 min to disperse the cells.

5. Alternatively, the SS may be placed in a small, sterile beaker, immersed in sterile process water, and the beaker placed in a waterbath sonicator to displace the cells from the surface. It is advisable to initially determine the optimal time of sonication in order to optimize recovery of cells.

6. Viable bacteria may also be assayed by aseptically scraping the stud surface with a sterile scalpel blade; the scrapings and the stud are then placed in sterile phosphate buffer and sonicated for 30-60 seconds. An aliquot of the

homogenate may then be plated on agar growth medium or filtered through a membrane for bacterial enumeration. The following equation may be used:

No. viable organisms = CFU/plate x dil. factor x 10

Example:

A 20 cm^2 surface area yielded 220 colonies on 10^{-6} plate

$220 \times 10^6 \times 10 = 2.2 \times 10^9$ cfu/coupon

2.2×10^9 cfu/coupon /20 cm^2 = 1.1×10^8 cfu/cm^2

Remember to take into account the volume of liquid in which the microorganisms are dispersed prior to serial dilution since this must be considered a dilution. Coupons can also be removed at timed intervals to estimate colonization rates.

Plate Count Methods

Pour Plate

The pour plate method involves small aliquot additions of homogenous samples (or serial dilutions thereof) into sterile petri plates, aseptic pouring of sterile molten agar into the plates, careful mixing, and enumeration of colonies that develop within the agar after an incubation period (usually 24-72 h) at a constant temperature. This general technique is used for the Standard Plate Count Method.

The layered pour plate method is a modification

of the above procedure. Layered pour plates are a small volume (3 ml) mixture of molten agar and cell suspension which is poured onto a solid minimal agar layer in a sterile petri plate. Advantages of layered pour plates include a) the colonies are easier to enumerate because of the thinner agar layer (compared to the traditional pour plate), b) the minimal agar layer can be stored in petri plates for extended periods for use as required, and c) flexibility in choice of nutrient suppliments for the molten agar overlay. Many researchers recommend the layered pour plate method in preference to the traditional pour plate or the spread plate techniques for colony counting.

Procedure

1. Prepare solid agar base. Pour sterile 1.5% w/v molten agar into sterile petri plates on a flat surface such that the agar layer is about 0.5 cm deep. After allowing the agar to harden, the plates can be used immediately or stored refrigerated until needed. Refrigerated agar should be warmed to room temperature before use.

2. Prepare soft agar overlay. Add about 3 ml soft Plate Count Agar (Tryptone Glucose Yeast Agar) to 13 x 100 mm screw-capped tubes. Sterilize the tubes by autoclaving. The tubes can be used immediately or stored refrigerated until needed. Before use, place the tubes into a 45°C water bath until temperature equilibrium is established; the agar will remain molten at 45°C for several hours.

3. Plate sample and dilutions of sample. Make 1/10 serial dilutions (usually 10^{-1} to 10^{-4}) of the sample in sterile buffer (0.5 M KH_2PO_4, pH 7.2). Add an 0.1 ml aliquot of sample or dilution to

the inside dry lip of a sterile soft agar overlay tube. Pour the soft agar onto the solid agar base in the petri plate such that the sample aliquot is quantitatively transferred with the overlay. Gently swirl the plate to evenly distribute the overlay. Assays should be performed at least in duplicate; triplicate assays are preferred. Allow plates to harden. Incubate at a temperature that corresponds to the water system from which the sample was obtained.

4. Count the colonies after a predetermined incubation period (usually 24-48 hrs.). Calculate the number of colony forming units per ml (CFU/ml) in the undiluted sample.

5. It is often helpful to present the data in logarithmic form. Determine $\log_{10}$ (CFU/ml) for each assay. Calculate the average $\log_{10}$ (CFU/ml) ± 1 standard deviation from duplicate or, preferably, triplicate plates.

Spread Plate

The spread plate method for enumerating viable microorganisms is preferred over the pour plate method. Inoculation of environmental bacteria into the 45°C molten agar employed with the pour plate methods inactivates many viable organisms which would otherwise form visible colonies. The spread plate method involves inoculation of a small sample volume or dilution thereof onto the surface of solidified agar medium at ambient temperature. The inoculum is then distributed over the agar surface using a glass or Teflon spreader.

Sample Collection and Processing

Materials

1. Suitable agar medium (0.3% TSA, Sabouraud Dextrose Agar, etc.) is sterilized and poured into sterile glass or plastic 100 mm petri dishes to a depth of about 0.3 cm. Approximately 18 ml of agar provides a sufficient base. The poured plates should be allowed to dry prior to use. Most agar media may be stored at $4^{o}C$ for up to 4 weeks.

2. Dilution blanks, 9 ml sterile phosphate buffer or saline solutions in test tubes.

3. Graduated, sterile 1 ml pipettes.

4. Glass or Teflon spreader. Glass spreaders may be formed by fusing the ends of a Pasteur pipette, then bending it to form an L-shaped "hockey stick". A large paper clip can be fitted into a section of Teflon tubing, then bent into the "hockey stick" form.

5. Ethanol or isopropanol for flame sterilization.

6. Colony counter with hand tally counter.

Procedure

1. Prepare 10-fold dilutions of the sample such that an 0.1 inoculum can be expected to yield between 30 and 300 colonies. For example, if the sample is expected to contain 10^6 cells/ml, five 10-fold sample dilutions should be made. The 10^3 through 10^6 dilutions would then be plated.

2. Pipette 0.1 ml of each sample dilution to be plated to the agar surface.

3. Dip the spreading surface into alcohol, shake off the excess fluid, then ignite. Allow the spreader to cool for a few seconds.

4. Spread the inoculum evenly across the agar surface. The inoculum should be absorbed into the agar within about 30 seconds; inadequately dried agar plates will not readily absorb inocula.

5. Incubate the labeled, inverted plates in an incubator at the appropriate temperature for 48-96 h.

6. The resulting colonies should be counted on a colony counter.

Data Evaluation

1. Calculate the number of viable organisms in the sample by the following formula:

$$\text{cells/ml} = \frac{\text{colonies counted}}{\text{dilution factor X 0.1}}$$

Membrane Filtration

Enumeration of microorganisms by membrane filtration requires the retention of viable cells on a presterilized membrane followed by incubation of the membrane on sterile, solid, nutrient-rich agar in petri plates. The nutrients in the agar layer diffuse through the porous membrane filter and provide for colony formation; colony enumeration is performed after an incubation period (usually 24-72 h) at a constant temperature. Advantages of the method include: a) large volumes of sample can be filtered;

thus the sensitivity of the technique is very good, b) versatility in nutrient supplements for the agar support layer and, c) the technique is simple. Membrane filtration is used most often in studies for which low concentrations of bacterial cells are anticipated but high cell concentrations can also be enumerated by making serial dilutions of the sample.

Preparation of Vacuum Filtration Unit

Use a glass filter holder assembly designed such that the membrane filter can be held in place securely. The most common filtration units consist of a funnel fastened by a clamping device to a fritted porous receptacle. Wrap the apparatus and sterilize by autoclaving. Use gridded, presterilized 47 mm diameter membrane filters.

Preparation of Agar Plates

Prepare petri plates of Plate Count Agar (Tryptone Glucose Yeast Agar); about 10-15 mls per plate is sufficient. Store the plates refrigerated until needed.

Sample Filtration

Select sample sizes according to the expected viable cell density. The number of colonies that develop on the membrane should be between 50 and 200. If high cell densities are expected (>200 CFU/ml), make 1:10 serial dilutions (usually 10^{-1} to 10^{-4}) of the sample in sterile buffer (0.5 M KH_2PO_4, pH 7.2). Use flame sterilized forceps to place the membrane filter in the filtration assembly. Add sample, filter by vacuum (approx 15 psi), aseptically remove filter, and place it (grid side up) on the agar plate.

Enumeration

Count the colonies after a predetermined incubation period (usually 24-48 hrs.) at the process water temperature. Calculate the number of colony forming units per ml (CFU/ml) in the undiluted sample.

Data Reduction

It is often helpful to present the data in logarithmic form. Determine $\log_{10}$ (CFU/ml) for each assay. Calculate the average $\log_{10}$ (CFU/ml) ± 1 standard deviation.

References

1. American Public Health Association. 1985. Standard methods for the examination of water and wastewater, 16th ed. American Public Health Association, Washington, D.C.

2. American Society for Microbiology. 1981. Manual of methods for general bacteriology, ASM, Washington, D.C.

Direct Counting

Direct enumeration is used to obtain total counts of microorganisms present in water. Depending upon the concentration of microorganisms present, either an on-line, large volume sampling or a small volume, vacuum filtration sampling is performed. This is followed by staining with a nucleic acid stain specific for microbial cells (acridine orange) and examination under an epifluorescent microscope. Epifluorescent microscopic counts yield results which may be several orders of magnitude greater than those obtained by cultural methodologies. The counts obtained are composites of viable and culturable, viable and non-culturable, aerobic, anaerobic, and non-viable microbial cells. This method does not allow differentiation of viable from non-viable cells, nor can it be used for species identification.

Materials and Reagents

1. 25 mm on-line stainless steel high pressure sampler (Nuclepore Corp., Pleasanton, CA or equivalent).

2. 25 mm glass microfiltration apparatus (Nuclepore Corp., Pleasanton, CA or equivalent).

3. Epifluorescent microscope: 100x oil immersion objective with a n.a. of 1.25, 10X eyepieces, high pressure mercury lamp (100 W), with a KV418 + BG12 exciter filter, 500 nm dichroic filter and DG515 barrier filter (American Optical, Buffalo or equivalent).

4. Vacuum pump, filtering flask/trap, tubing.

5. Sample collection container, graduated.

6. Other supplies: flat tipped forceps, glass slides, coverslips, pipets, screw-top test tubes.

7. 25 mm dia., 0.2 μm pore size, predarkened, polycarbonate membrane filters (Nuclepore Corp., Pleasanton, CA or equivalent).

8. 25 mm dia., 5.0 μm pore size, cellulosic support membranes (Nuclepore Corp., Pleasanton, CA or equivalent).

9. Fixative: 5% glutaraldehyde in phosphate buffer.

10. Fluorochrome: 0.1% (w/v) acridine orange in pH 7, 0.1 M phosphate buffer, biological grade (Sigma Chemical Co., St. Louis, MO or equivalent).

11. Low fluorescing immersion oil (Cargille Type B or equivalent).

12. Eyepiece counting graticule, (American Optical, Buffalo or equivalent).

Procedure (water sample)

1. Preclean the on-line sampler or microfiltration apparatus in an appropriate cleaning solution.

2. Prerinse the on-line sampler or microfiltration apparatus with the process water. Controls should be run to determine background levels.

3. Assemble the on-line sampler by placing the 0.2 μm filter on top of the 5.0 μm support and attaching the sampler to the piping system (purified water systems typically contain relatively low numbers of planktonic microorganisms, thus necessitating the use of a large volume sample apparatus). Samples of 500 ml to several liters can be obtained by this method. The volume sampled is determined by collection of the filtrate in a carboy or other graduated sample container. Smaller samples from waters containing relatively high numbers of microorganisms (cooling, municipal, etc.) may be filtered off-line through the vacuum filtration apparatus. The 0.2 μm membrane is placed directly on the sintered glass support in preparation for sampling. A vacuum of <100 mm Hg should be employed. Samples may be fixed with glutaralde-

hyde (0.05% final concentration) and stored prior to processing. The sample volume required is dependent upon the bioburden present (see step 7, below).

4. After the sample has passed through the 0.2 μm membrane, carefully add approximately 1 ml of the fluorochrome to the membrane surface and let stand for 2 minutes.

5. Remove the residual stain from the membrane surface by applying a vacuum. Remove the membrane from the sampler or microfiltration apparatus and allow to air dry at room temperature (1-2 min).

6. Place 1 drop of low fluorescing immersion oil on the surface of a clean, glass microscope slide, place the stained membrane (side containing trapped microorgisms should be up) on the oil, add 1 drop of oil to the surface of the membrane, then place a clean glass coverslip on top of the membrane. The prepared sample can be stored in a cool, dark area for several weeks if it is not possible to perform the enumeration step immediately.

7. Place the prepared slide under the 100X oil immersion objective of an epifluorescent microscope and count all cells present, regardless of their color. Cells in clusters should be counted individually. If the number of cells per viewing grid is too low for precise counting, resample with a larger sample volume; if there are too many cells present, resample with a smaller sample volume. Between 20 and 200 viewing grids should be counted, depending upon the concentration of cells present.

8. Total numbers of cells present in a 100 ml sample are calculated from the formula shown below:

$$N = \frac{a \times n \times 0.1}{m \times v}$$

where,

N = total number of cells/100 ml
a = effective filtration area in mm^2
n = mean count of cells present per area of filter delineated by the eyepiece counting graticule
m = area enclosed by the eyepiece reticule at magnification in mm^2
v = sample volume, liters

Example: A 50 ml sample is filtered through a membrane with an effective filtration area of 300 mm^2. An optical graticule with an area of 0.0025 mm^2 @ 1000X is used for counting. Fifty fields are counted, yielding an average count of 2.5 cells/field.

$$N = \frac{300 \times 2.5 \times 0.1}{0.0025 \times 0.05}$$

$$= 6.0 \times 10^5 \text{ cells/100 ml}$$

Procedure (surface sample)

1. Fouled sample surfaces can be examined directly using the epifluorescent direct-count assay.

Sample Collection and Processing

The sampled surfaces should be carefully rinsed with prefiltered water, then stained with sufficient acridine orange (or DAPI) to cover the entire surface.

2. Following a two-minute staining period, the surface is allowed to air dry. To facilitate viewing, a coverslip may be placed on a drop of immersion oil over the stud surface.

3. Calculate the density of organisms on the sampled surfaces using the following formula:

$$N = n/m$$

where, N = total cells/cm^2
n = mean cell count per area of surface delineated by the eyepiece counting reticule
m = area enclosed by the eyepiece graticule at magnification in cm^2

References

1. American Society for Testing and Materials. 1985. Enumeration of aquatic bacteria by epifluorescence microscopy counting procedure. p. 632-635. In Water, Vol.11.02, D4455-85, American Society for Testing and Materials, Philadelphia.

2. Geesey, G.G., R. Mutch, J.W. Costerton, and R.B. Green. 1978. Sessile bacteria: an important component of the microbial population in small mountain streams. Limnol. Oceanogr. 23:1214-1223.

3. Mittelman, M.W., G.G. Geesey, and R. Hite. 1983. Epifluorescence microscopy: a rapid method for enumerating bacteria in ultrapure water

systems. Microcontamination 1:(2)32-37;52.

Use of Scanning Electron Microscopy for Biofouling Evaluation

Scanning electron microscopy (SEM) has become an indispensable tool for evaluating the superficial fine structure of sessile microbial populations. The technique may be used in a strictly qualitative fashion to determine the morphological diversity of microorganisms comprising the biofilm, or it may be employed quantitatively to estimate the numbers and distributions of attached or free bacteria, or the overall extent of surface biofouling. When applied in conjunction with energy dispersive x-ray microanalysis (EDX), a complete semiquantitative elemental analysis of the biologically-fouled surface may be obtained, thereby providing valuable information concerning the type and degree of chemical fouling within the sample.

Because of the extremely fragile nature of the many bacteria and other biofouling microorganisms, the intracellular and extracellular macromolecules (proteins, nucleic acids, lipids, and carbohydrates) responsible for maintaining cellular architecture and biofilm structure must be preserved by crosslinking them *in situ* using one or more chemical fixatives prior to examination by SEM [4-6]. The biofilm sample preparation method outlined below represents only one of the more popular preparative schedules that have been devised to accomodate biofilm specimens. Many other similar procedures and variations are currently in practice, some of which are described in detail in the references cited at the end of this section.

Sample Collection and Processing

Water Samples

Water samples are chemically fixed for periods of 24-96 h at 4°C using a final concentration of 2.5% (wt/vol) electron-microscopy grade glutaraldehyde (Pelco, Inc., Tustin, CA). Appropriate aliquots of the fixed samples (from 10-400 ml, depending on the amount of suspended particulate matter present) are filtered through a 0.2 μm pore size Nuclepore membrane filter. Bacteria and other particulate matter retained on the membrane surface are washed with 3 consecutive changes of distilled water (10 ml each). The particulates are then dehydrated in an increasing ethanol concentration series (30, 50, 75, 85, 95, 100% x 3, three min. each), followed by a 1:1 mixture of absolute ethanol:Freon 113 for 3 min., and finally three changes of absolute Freon 113. The membranes are air dried under vacuum directly from the final Freon 113 solution, removed from the filtration apparatus, cut into 1 cm squares, and mounted on aluminum stubs using conducting silver paint.

Biofilms

The sampling device containing the attached biofilm is carefully removed from the system minimizing mechanical and fluid disturbance to the sessile microorganisms. Before any part of the biofilm is subject to desiccation, the sample is submerged in an ice-cold solution of glutaraldehyde (2.5% wt/vol) in 50 mM monobasic potassium phosphate buffer and adjusted to the system pH with 1.0 N KOH or HCl. Following overnight fixation in a refrigerator, the glutaraldehyde solution is removed by immersing the biofilm in 3 consecutive changes of distilled water at room temperature. If the biofilm is dense, it is advisable to allow the sample to soak for at least 5-10 min at each rinse step. Care must be exercised at this and later stages not to disturb the attached biofilm by rapid or violent mixing of the rinse solutions. The sample is dehydrated in a graded ethanol series as described above except that the biofilm is immersed for 5-10 min in each solution. (A word of caution: Some batches of ethanol and Freon 113 may require prior filtration through an 0.2 μm pore size Nuclepore membrane filter to remove suspended particulate matter. In addition, water used for preparing all solutions must be particle-free). The biofilm may be stored indefinitely at 4°C once it has been transferred to 100% Freon 113.

If the biofilm is to be critically-point dried, Freon 113 is used as the intermediate transitional fluid in the drying apparatus [3]. In many instances, however, it is not absolutely necessary to employ critical-point drying and the biofilm can be air-dried directly from the 100% Freon 113 solution without severe distortion of microbial surface structure. In the event that the sample is air dried, it is advisable to further transfer the biofilm through 1 or 2 additional changes of 100% Freon 113 to ensure that all traces of ethanol and moisture have been removed from the sample. After the final transfer, uniform and very rapid drying of the biofilm surface is essential for the success of this step. The air-dried sample is mounted on an aluminum stub with conducting silver paint, colloidal graphite, or some other appropriate adhesive.

Sputter Coating

The stubs containing either the membrane-filter or biofilm samples are coated with a film (about 200 angstroms thick) of gold:palladium (60:40) using a vacuum coating device. At this point the specimens are ready to examine in a SEM. Photographic records of areas of the specimen should be made.

References

1. Cohen, A.L., D.P. Marlow, and G.E. Garner. 1968. A rapid critical point method using fluorocarbons ("freons") as intermediate transitional fluids. J. Microsc. (Paris) 7:331-342.

2. Gabriel, B.L. 1982. Biological scanning electron microscopy. Van Nostrand Reinhold Co., New York.

3. Hayat, M.A. 1974-1976. Principles and techniques of scanning electron microscopy, Vols. 1-5, Van Nostrand Reinhold Co., New York.

4. Hayat, M.A. 1982. Fixation for electron microscopy. Academic Press, London.

5. Postek, M.T., K.S. Howard, A.A. Johnson, and K.L. McMichael. 1980. Scanning electron microscopy, A student's handbook, Ladd Research Industries, Inc.

CHAPTER 11. Microbial Identification

The subject of microbial identification is of a broad and diverse nature. Although a large body of information is available on the isolation and identification of medically significant microorganisms, relatively little has been published concerning identification of environmental bacteria. A detailed discussion of microbial identification procedures is beyond the scope of this text. The following information is presented to provide interested readers with applicable references for further study.

Bacteria

The unique cultural and biochemical characteristics of waterborne industrial bacteria create a number of difficulties in isolation and identification. Often, a primary environmental isolate will be lost after one subculturing. In addition, many of these organisms require extended incubation times for colony formation. Media and test reagents must be specific for this group. Bacteria associated with industrial waters range in size from somewhat less than 0.1 μm in diameter to over 100 μm in length for some of the filamentous forms. The isolation and test protocol shown in Figure 1 is designed to recover and identify a number of the more common waterborne industrial isolates. Selected media and diagnostic reagents required for isolation are referenced in Table 1. Table 2 lists some of the diagnostic reactions associated with the more common non-fermentative bacilli often associated with industrial waters. Interested readers may also consult the following reference sources for further information concerning the identification of bacteria associated with industrial waters:

1. American Public Health Association. 1985. Standard methods for the examination of water and wastewater, 16th ed., APHA, Washington, D.C.

2. Greenwood, J.R. 1983. A new approach to identifying nonfermentative bacilli. Particulate and Microb. Contamin. Jul/Aug:51-54.

3. Krieg, N.R. and J.G. Holt (ed.). 1984. Bergey's manual of systematic bacteriology, Vol. 1., Williams & Wilkins, Baltimore.

4. McCoy, J.W. 1980. The microbiology of cooling water. Chemical Publishing, New York.

5. Sneath, P.H.A., N.S. Mair, M.E. Sharpe, and J.G. Holt (ed.). 1986. Bergey's manual of systematic bacteriology, Vol. 2, Williams & Wilkins, Baltimore.

6. Stanier, R.Y., E.A. Adelberg, J.L. Ingraham, and M.L. Wheelis. 1979. Introduction to the microbial world, Prentice-Hall, Englewood Cliffs, NJ.

Fungi

This group consists of the molds and yeasts. Aquatic molds are referred to as aquatic *Phycomycetes*. They are often found associated with decaying plant and/or animal matter. The vegetative structure of molds is typically a sac, 100 µm or more in diameter, containing the reproductive cells (spores). The spores are typically spherical and 5-10 µm in diameter. Molds are typically identified using morphological criteria. Yeasts are unicellular fungi that exist as small (1-10 μm in diameter), spherical cells that reproduce by forming buds. Mycelia, which are typical of aquatic Phycomycetes, are absent in this group. Yeasts are often identified using biochemical tests similar to those employed for bacterial identifications. Figure 2 provides an outline for primary isolation and differentiation of molds and yeasts. The following reference provides a key to the identification of this group:

Koneman, E.W., G.D. Roberts, and S.S. Wright. 1978. Practical laboratory mycology, 2nd ed., Williams & Wilkins, Baltimore.

Algae

As with molds and yeasts, identification of algae is generally based upon morphological, rather than biochemical characteristics. Most algae contain the green pigment chlorophyll, that is characteristic of plants. Excellent keys to the identification of algae and cyanobacteria (blue-green algae) may be found in the following references:

1. American Public Health Association. 1985. Standard methods for the examination of water and wastewater, 16th ed., APHA, Washington, D.C.

2. Bold, H.C. 1973. Morphology of plants. Harper & Row, New York.

3. Benson, H.J. 1979. Microbiological applications (Laboratory manual), W.C. Brown, Dubuque, Iowa.

TABLE 1. Media and reagents used in the isolation and identification of bacteria, fungi, and algae

Medium or Reagent	Application	Supplier
trypticase soy agar/broth (TSA/TSB)	bacterial isolation	1, 2, 3
peptone agar/broth	bacterial isolation	1, 2, 3
Sabouraud dextrose agar	fungal isolation	1, 2
R2A agar	bacterial isolation	1
API Systems	bacterial/yeast identification	4
N/F Systems	bacterial identification	5
oxidase swabs	bacterial identification	6
blood agar	bacterial/fungal isolation	2
Enterotube Systems	bacterial identification	7
starch agar	bacterial identification	1

1. Difco Laboratories, Detroit, MI
2. BBL Laboratories, Cockeysville, MD
3. Scott Laboratories, Carson, CA
4. API Systems, Los Angeles, CA
5. Corning Products, New York
6. Cal-Scott, North Hollywood, CA
7. Roche Diagnostics, Morris Plains, NJ

TABLE 2. Reactions of some common purified water system non-fermentative bacilli (Gram - rods not fermenting glucose).

Organism	Test[1]							
	1	2	3	4	5	6	7	8
Pseudomonas aeruginosa	+	+	V	-	-	+	-	+
P. putida	+	+	-	-	-	+	-	-
P. fluorescens	+	+	-	-	-	+	-	-
P. cepacia	+	-	V	+	+	+	-	V
P. stutzeri	+	-	V	-	-	+	-	-
P. acidovorams	+	-	+	-	-	+	-	+
P. maltophilia	-	-	-	+	-	+	-	-
Alcaligenes odorans	+	-	+	-	-	+	-	+
Flavobacterium spp.	+	-	-	-	-	-	+	-
Acinetobacter anitratus	-	-	-	-	+	-	-	-
A. lwoffi	-	-	-	-	-	-	-	-
CDC group Ve-1	-	-	-	V	-	+	-	-

+, positive reaction
-, negative reaction
V, variable reaction

[1]Tests: 1. oxidase 2. fluorescence 3. denitrification 4. lysine decarboxylation 5. lactose oxidation 6. motility 7. indole hydrolysis 8. acetamide hydrolysis

Microbial Identification

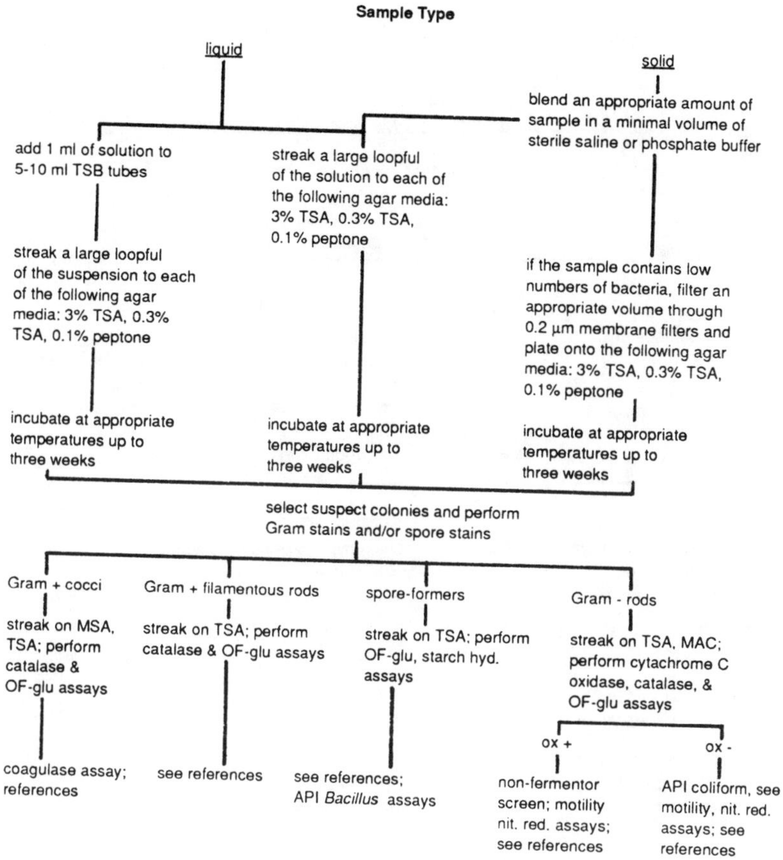

Figure 1. Flow-chart for the isolation and preliminary identification of waterborne, industrial bacteria.

Initial Isolation
(filter surface, scraping, sediments, overlying water)

Sabouraud dextrose agar, blood agar with antibiotics

Macroscopic Examination
(colonial morphology)

Molds

filamentous mycelia

Yeasts

non filamentous, raised, smooth, mucoid surface; pigmented-non pigmented

Microscopic Examination
(staining with lactophenol cotton-blue)

multicellular
septate or non septate hyphae
sexual spores
asexual spores

unicellular
lacking hyphae or pseudohyphae
sexual spores
asexual spores

Biochemical Assays

sugar fermentations
nitrate reduction
urea hydrolysis

sugar fermentations
nitrate reduction
urea hydrolysis

Figure 2. Primary isolation and differentiation of molds and yeasts.

CHAPTER 12. BIOCHEMICAL MONITORING OF FOULING

Biofouling may be detected and quantitatively evaluated by assays developed to measure specific components that are found in microbial cells. Some components, such as adenosine triphosphate (ATP), protein, and carbohydrate, are found in all living cells while others are associated only with specific groups of microorganisms such as lipopolysaccharide in Gram-negative bacteria. Although the quantities of these components may vary within microbial cells, they have often been used as a measure of microbial biomass. In most industrial water systems, the presence of these cellular components is generally good evidence and a useful measure of the extent of biofouling. The assays outlined below may be applied to both planktonic water samples as well as diluted homogenates of sessile biofilms.

Adenosine Triphosphate (ATP) Assay

Reagents and Materials

1. Purified Luciferin-Luciferase reagent (Sigma Chemical Co, St. Louis, MO, or equivalent).

2. Standard ATP, obtained from Sigma Chemical Company, St. Louis, MO. Equine muscle disodium salt, crystalline form. Preweighed vials contain 30 mg ATP salt. Store at -20°C.

3. TRIS buffer, 0.02 M TRIS, pH 7.75.

4. Filter apparatus capable of processing up to 10 l pyrogen free (pf) water through a 47 mm diameter membrane filter.

5. Sartorius 0.45 or 0.22 μm dia. cellulose nitrate membrane, 47 mm diameter, Sartorius, Hayward, CA. (or equivalent).

6. Glass test tubes 20 x 150 mm.

7. Hot plate for boiling TRIS buffer.

8. Luminometer; a partial list of luminometer suppliers:

   Aminco, Silver Spring, MD
   Berthold Analytical Instuments, Nashua, NH
   LKB, Gaithersburg, MD
   LUMAC, Schaesberg, The Netherlands
   Packard, Downers Grove, IL (now supplied by Los Alamos Diagnostics, Los Alamos, NM)
   3M, St. Paul, MN
   Turner Designs, Mt. View, CA

Preparation of ATP Standard

1. Add 24.5 ml of water to 30 mg of preweighed Sigma ATP. Since the molecular weight (MW) of pure ATP is 507.2 and the MW of the Sigma salt is 614.2 and the purity of the Sigma product is 99%:

   614.2 x 30 mg x 0.99 = 24.5 mg

   24.5 mg of pure ATP is contained in a 30 mg vial of the Sigma product. By adding 24.5 ml pf distilled water, a solution containing 1000 micrograms of ATP/ml is obtained (Employ this approach when using ATP obtained from sources other than Sigma). Dispense solution in 1.0 ml aliquots and store at -20°C.

2. Dilute 1000 μg/ml stock ATP solution serially 1:10

in sterile, distilled water to 10^{-6} μg/ml. It is necessary to prepare fresh standard ATP from frozen stock for each assay.

Preparation of Luciferin-Luciferase

1. Most commercial preparations are complete and require only reconstitution with distilled water or a supplied reconstitution buffer. Follow directions as supplied with the reagent. It should be noted that immediately following reconstitution, the luciferase mixture may spontaneously emit light due to traces of ATP left during the purification procedure. It is advisable, therefore, to incubate the mixture at room temperature for 30-60 min to reduce the level of background light.

2. Follow directions from supplier for use and storage of reconstituted reagent.

Specimen Collection and Preparation

1. Filter a volume of water through the 0.45 μm filter (1-10 l). Volume filtered is dependent on the microbial content.

2. Remove wet filter with forceps and immediately place in boiling TRIS (10.0 ml in a 20 x 150 mm test tube). Continue boiling for 5 min. Remove tube from heat and place on ice to cool.

Assay Procedure

1. Place sample or dilution of sample into luminometer cuvette (a typical volume is 0.5 ml).

2. Add 0.5 ml of Luciferin-Luciferase reagent (note: some luminometers may require different volumes, follow manufacturer's directions).

3. Immediately place cuvette in luminometer and read results.

Calculations

1. ATP detection is usually in the range of 10^{-5} to 10^{-1} micrograms of ATP per ml sample. Readouts on most instruments are in relative units based on light emitted. These units can be converted to ATP through the use of standard ATP and the construction of a standard curve; i.e., light units vs. ATP concentration.

2. Subtract background light emission from readings before calculating unknowns or constructing standard curves.

3. For some applications, standard curves can be constructed directly for the parameter to be measured; e.g., light units vs. bacterial direct count, etc.

References

1. Deming, J.W., G.L. Picciolo, and E.W. Chappelle. 1979. Important factors in adenosine triphosphate determinations using firefly luciferase: Applicability of the assay to studies of native aquatic bacteria. p. 89-98. In J.W. Costerton and R.R. Colwell (ed.), Native aquatic bacteria: enumeration, activity, and ecology. ASTM, Philadelphia.

2. Geesey, G.G. and J.W. Costerton. 1979. Bacterial biomass determinations in a silt-laden river: Comparison of direct count epifluorescence microscopy and extractable adenosine triphosphate techniques. pp. 117-127. In J.W. Costerton and R.R. Colwell (ed.), Native aquatic bacteria: enumera-

tion, activity, and ecology. ASTM, Philadelphia.

3. Holm-Hansen, O. 1973. Determination of total microbial biomass by measurement of ATP. pp. 73-89. *In* L.H. Stevenson and R.R. Colwell (ed.), Estuarine microbial ecology. University of South Carolina Press, Columbia, South Carolina.

Limulus Amebocyte Lysate (LAL) Assay

Reagents and Materials

1. Freeze-dried LAL, 5.0 ml/vial, (2.5 ml/vial for chromogenic assay). The folowing companies supply reagents required for the LAL assay: Associates of Cape Cod, Inc., Woods Hole, MA; Haemachem, Inc., St. Louis, MO; Marine Biologicals, Inc., Marmora, NJ; Whitaker M.A. Bioproducts, Walkersville, MD.

2. Standard endotoxin.

3. LAL reagent water.

4. Reaction tubes. 10x75 mm soda lime glass for the gel-clot assay or 12x75 mm borosilicate glass for the turbidimetric or chromogenic tests.

5. Hot air oven with 180° or 250°C capacity for depyrogenation of glassware.

6. Water bath without agitator or block heater capable of maintaining 37°C (plus or minus 0.5^{o}C) for incubation of gel-clot and chromogenic tests or a LAL-4000/5000 (Associates of Cape Cod, Inc.) or a kinetic microplate reader for the turbidimetric assay.

7. Sterile, disposable 1, 5, and 10 ml pipettes,

and/or automatic pipettes equipped with pyrogen-free tips.

8. Test tube rack for holding reaction tubes.

9. Vortex type mixer.

10. Parafilm (3M Company, St. Paul, MN). The side contacting the backing paper is usually pyrogen-free.

11. Pyrogen-free test tubes for making dilutions.

Preparation of standard endotoxin

1. Reconstitute dry standard with pyrogen-free water.

2. Make a series of dilutions to cover range of LAL sensitivity.

3. Be sure to thoroughly mix endotoxin solutions during the dilution process and immediately before testing with LAL.

Preparation of LAL

1. Remove the stopper and discard. Reconstitute with 5.0 ml (or 3.5 ml for chromogenic reagent) of LAL reagent water. Swirl gently to dissolve--do not vortex. Do not replace stopper--use parafilm. Store LAL on ice prior to use.

2. Reconstituted LAL is slightly cloudy and opalescent. If gross particulates or flocs are present or solution is distinctly yellow, do not use.

Biochemical Monitoring

Specimen collection and preparation

1. Specimens for analysis must be collected aseptically in pyrogen-free containers.

2. Adjust the pH of sample to 6-7.5 using sterile, pyrogen-free buffer or acid or base. DO NOT ADJUST THE PH OF WATER OR SALINE.

3. Make an appropriate series of two-fold dilutions of the test sample

The gel-clot assay

Test procedure

1. Add 0.1 ml of each dilution of sample to each of a series of reaction tubes.

2. Add 0.1 ml of reconstituted LAL to each tube, starting with the blank or negative control and ending with the highest endotoxin concentration. Use a fresh pipette for each entry into the LAL container. Shake the rack of tubes vigorously for 30 to 60 seconds to insure thorough mixing of sample and reagent. Failure to mix adequately is a common problem encountered in performing satisfactory LAL assays. CONSISTENT technique is necessary to obtain reproducible results.

3. Place the reaction tubes in a 37°C water or dry bath for 60 plus or minus 2 min. Timing of the reaction is critical. If large numbers of samples are to be tested in parallel, the reactions should be started at appropriate intervals so as to

permit reading of each test within the prescribed time limit. The gel formation is delicate and may be irreversibly terminated if the tubes are handled or are subjected to agitation or vibration. Do not use a circulating water bath. Submerge the tubes above the level of the reaction mixture but not so high that they float.

The endotoxin standard

1. For each day's work prepare a fresh set of dilutions from stock endotoxin solution. Test twofold dilutions that bracket the sensitivity of the LAL.

2. Transfer 0.1 ml of each concentration of endotoxin to separate reaction tubes.

3. Starting with the negative control, add 0.1 ml of reconstituted LAL and mix as described above.

4. Place tubes in the incubator as described above.

Performing the assay: Controls

1. Negative control. Add 0.1 ml of LAL reagent water to one tube to serve as a blank or negative control.

2. Positive control. The standards as described above serve as positive controls.

3. Positive product control. When testing samples other than water a positive product control should be included. Add endotoxin to the sample under test such that the final

concentration in the sample is twice as great as the rated sensitivity of the LAL. This control alerts the user to possible interferring substances in the sample.

4. Add 0.1 mL of LAL to each of the control tubes and incubate as described above.

Reading results

1. To obtain consistent, reliable results, one must handle the reaction tubes carefully and uniformly in the reading process. Remove reaction tubes one at a time. Slowly invert each in one smooth motion. Do not pause halfway in the inversion. DO NOT WIPE TUBES OR BUMP THEM against the rack during removal.

2. A positive test is indicated by the formation of a gel which does not collapse when the tube is completely inverted. That is, if a sufficient concentration of endotoxin is present in the sample, a firm clot will be formed which will not break when tubes are carefully inverted.

Interpretation

1. The results of a representative lot of LAL with a labeled sensitivity of 0.25 EU/ml would appear as follows:

   0.5 EU/ml................(+)
   0.25 EU/ml...............(+)
   0.125 EU/ml..............(-)
   0.06 EU/ml...............(-)

   This particular lot of LAL would thus have a confirmed sensitivity of 0.25 EU/ml.

2. If negative control clots: water, glassware, or LAL is contaminated.

3. Positive control: standard values should be at least plus or minus two-fold of the labeled sensitivity; if not, check dilutions or for possible loss of endotoxin through adsorption to container walls.

4. Positive product control fails to clot: product is inhibitory to LAL and further dilution of sample is required.

5. To calculate the concentration of endotoxin in the test samples multiply the sensitivity of the LAL by the dilution factor of the highest dilution giving a positive test.

   Example: Highest dilution showing a positive test = 1:100, sensitivity of LAL = 0.25 EU/ml, sample contains 100 x 0.25 EU/ml or 25 EU/ml.

The Turbidimetric Assay

Test procedure

1. Add 0.4 ml of each sample dilution or endotoxin standard to a 12x75 mm borosilicate tube (if using the microplate method add 0.1 ml of sample to a well).

2. Add 0.1 ml of reconstituted LAL (formulated for turbidimetric assay) to the first tube. Vortex immediately and place in the LAL-4000/5000. Repeat this procedure for the remaining tubes. Refer to the LAL-

4000/5000 operations manual for specific instructions. When using the microplate method, RAPIDLY add the LAL to all wells containing sample. Use of a repeating pipette facilitates this procedure. Immediately following addition, mix the entire plate on a vibrator platform and place the plate in the incubator/reader.

The standard

1. Prepare four to six two-fold dilutions of endotoxin standards within the range of 0.001 to 100 EU/ml. Perform the test as described above.

2. To construct a standard curve, plot the log onset time against the log of the endotoxin concentration. (The onset time is the time required to reach a preset optical density, usually 0.02 in the LAL-4000/5000 and 0.05 in the microplate reader). The absolute value of the correlation coefficient (r) should always be equal to or greater than 0.980.

3. To calculate endotoxin values in test samples, enter the time of onset into the line equation of the standard and calculate the result. Computer programs available for both the LAL-4000/5000 and the microplate reader do this automatically.

The Chromogenic Assay

Test procedure

1. Place 10x75 mm test tubes in a test tube rack.

2. Aseptically dispense 0.1 ml of LRW into one test tube. This serves as a negative control and can be used to set the spectrophotometer to zero.

3. Dispense 0.1 ml of the standards (1.0, 0.5, 0.25, and 0.125 EU/ml) into each of four test tubes. These tubes will be used to construct a linear regression curve.

4. Dispense 0.1 ml of sample into a tube.

5. Aseptically dispense 0.1 ml of LAL reagent into each tube.

6. Remove tubes from rack, mix thoroughly, and place into a 37°C $\pm$1°C for the appropriate incubation time (see determination of incubation time below).

7. At the end of the incubation period $\pm$10 sec, add 0.05 ml of 50% acetic acid to each tube to stop the reaction. The reaction tubes must be stopped rapidly using a repeating pipetter so the time from the first tube to the last tube is just a few seconds.

8. Read the test results in a spectrophotometer at 405 nm. The spectrophotometer must be zeroed on the negative control.

Determination of incubation time

The time of reaction is directly related to the optical density (O.D.) of the reactants. The greater the time of reaction, the greater the O.D., or conversely, the less the reaction time, the lower the O.D. for a given concentration of endotoxin. Other parameters also affect the O.D. Some of these are temperature of the reaction mixture, length of the light path, etc. By holding these additional factors constant, the O.D. can be controlled by varying the incubation time. A suggested incubation time is provided for each lot of LAL reagent for a standard in the range of 1.0 to 0.125 EU/ml.

References

1. Evans, T.M., J.E. Schillinger, and T.G. Stuart. 1978. Rapid determination of bacteriological water quality by using Limulus lysate. Appl. Environ. Microbiol. 35:376-382.

2. Novitsky, T.J., M.E. Dawson, and W.V. Collentro. 1985. A new approach for monitoring bacteria: The kinetic LAL assay. Ultrapure Water November/December p. 10-12.

3. Novitsky, T.J., J. Schmidt-Gengenbach, and J.F. Remillard. 1986. Factors affecting recovery of endotoxin adsorbed to container surfaces. J. Parent. Sci. Technol. 40:284-286.

4. Remillard, J.F., M.C. Gould, P.F. Roslansky, and T.J. Novitsky. 1987. Quantitation of endotoxin in products using the LAL kinetic turbidimetric assay. p. 197-210. In S.W. Watson, J. Levin, and T.J. Novitsky (ed.), Detection of bacterial endotoxins with the Limulus amebocyte lysate test, Alan R. Liss, Inc., New York.

5. Watson, S.W., T.J. Novitsky, H.L. Quinby, and F.W. Valois. 1977. Determination of bacterial number and biomass in the marine environment. Appl. Environ. Microbiol. 33:940-946.

Lowry Protein Assay

Reagents and Materials

1. Reagent A: Prepare a 2% Na_2CO_3 solution in 0.1 N NaOH. Dissolve 4.0 g NaOH in distilled water to a final volume of 1 liter. Dissolve 20 g Na_2CO_3 in 980 ml NaOH solution to give Reagent A.

2. Reagent B: Prepare a 1% $CuSO_4$ solution by dissolving 5 g $CuSO_4$ in 495 ml of distilled water.

3. Reagent C: Prepare a 2% sodium tartrate solution by dissolving 10 g sodium tartrate in 490 ml distilled water.

4. Reagent D: Dilute Folin Reagent 1:1 with distilled water to achieve a 1 normal concentration.

Preparation of Standard

1. Dissolve 10 mg of bovine serum albumin in 100 ml of Reagent A to achieve a working stock concentration of 100 μg/ml.

2. Make dilutions of the stock solution according to Table 1.

Assay Procedure

1. Prepare fresh each day Reagent E by mixing 100 parts Reagent A with 1 part Reagent B and 1 part Reagent C.

2. Prepare one tube containing 0.5 ml of sample solution.

3. To each tube containing 0.5 ml of sample or standard solutions add 2.5 ml Reagent E, mix and let stand for 10 min.

4. Add rapidly 0.25 ml Reagent D to each tube and mix immediately.

5. Let all tubes stand for 30 min and then read absorbance at 550 nm.

6. Interpolate protein concentration of sample from standard curve by plotting protein concentration vs. absorbance.

TABLE 1. Protein standard dilution protocol

Tube	Concentration (μg/ml)	Microliters of standard solution	Microliters of distilled water
1	100	500	0
2	75	375	125
3	50	250	250
4	25	125	375
5	10	50	450
6	0	0	500

Reference

1. Lowry, O.H., N.J. Rosebrough, A.L. Farr, and R.J. Randall. 1951. Protein measurement with the Folin phenol reagent. J. Biol. Chem. 193:265-275.

Coomassie Brilliant Blue Protein Assay

This method is suitable for samples that contain cations such as potassium or sodium and carbohydrates such as sucrose. Strongly alkaline buffering agents, however, produce a small amount of interference while detergents give an excessive color interference.

Reagent and Materials

Coomassie Blue Reagent: Dissolve 100 mg Coomassie Brilliant Blue G-250 (Sigma Chemical Co.) in 50 ml 95% ethanol. Add 100 ml of 85% (w/v) phosphoric acid and then dilute the mixture to 1000 ml final volume with distilled water.

Standard Preparation

Bovine serum albumin is made up in distilled water to a working concentration of 100 μg/ml.

Assay Procedure

1. Pipet 0.1 ml of sample containing from 10-100 μg protein into 12 x 100 mm test tubes.

2. Pipet 0.1 ml of water or buffer (whatever was used to make up standard solutions) into test tubes.

3. Prepare a series of tubes containing different known amounts of bovine serum albumin according to Table 1. Distribute 0.1 ml of each diluted standard solution into test tubes.

4. Add 5 ml Coomassie Blue Reagent to the tubes and mix thoroughly on a vortex.

5. Read absorbance at 595 nm after 2 min but before 1 h using a spectrometer. The spectrometer is first zeroed using the blank.

6. Prepare a standard curve by plotting the amount of bovine serum albumin vs. absorbance. The amount of protein in the sample is then determined from the curve on the basis of the absorbance obtained. If the protein content is too high, appropriate dilutions of sample should be made before adding the Coomassie Blue Reagent.

7. A sample containing 5 ug protein/ml should give an absorbance of approximately 0.275.

Reference

1. Bradford, M.M. 1976. A rapid and sensitive method for the quantitation of microgram quantities of protein utilizing the principle of protein-dye binding. Anal. Biochem. 72:248-254.

Carbohydrate Assay

Reagents and Materials

1. 25 g of 2X distilled phenol in 500 ml distilled water (phenol reagent).

2. 2.5 g hydrazine sulfate in 500 ml concentrated sulfuric acid (sulfuric acid reagent).

3. 10 mg of glucose in 100 ml of distilled water (stock standard solution).

Sample Preparation

1. If the sample consists of a biofilm:

a. Quantitatively remove approximately 0.2 g of sample from test surface and immerse in a minimum volume of distilled water. Sonicate for 1-5 minutes to release biofilm into suspension, vortex, and then make 10-fold serial dilutions of the liquid phase.

b. Add 0.5 ml of phenol reagent to 0.5 ml of each sample dilution in a clean test tube.

2. If the sample consists of an aqueous suspension:

   a. Add 0.5 ml of phenol reagent to 0.5 ml of water sample.

Standard Preparation

1. Prepare different concentrations of the glucose standard according to Table 2.

2. Add 0.5 ml phenol reagent to each of the 6 standard solutions prepared above.

Assay Procedure

1. Add 2.5 ml sulfuric acid reagent as rapidly as possible to each tube containing standard solutions and sample dilutions. Mix each tube on a vortex mixer during acid addition. This step must be performed in a reproducible manner from one tube to the next in order to provide adequate analytical precision.

2. Allow tubes to stand for 1 h at room temperature to cool.

3. Measure absorbance of each sample in a spectro-

photometer at a wavelength setting of 490 nm, using a 1 cm pathlength cell. Zero the instrument first with the blank solution.

4. Measure the absorbance of each standard solution, then construct a standard curve to determine the carbohydrate concentration of each sample.

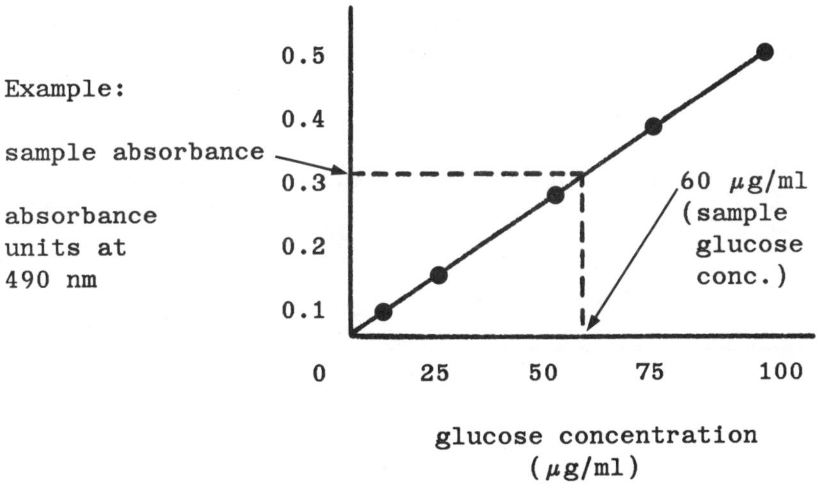

Calculations

1. Sample glucose concentration from standard curve is multiplied by the dilution factor (10^{-1} dilution = dilution factor of 10; 10^{-2} dilution = dilution factor of 10^2, etc.) to determine glucose concentration in the undiluted biofilm. This value is expressed in terms of the area of surface from which the biofilm was obtained.

2. Glucose concentration of aqueous sample is determined directly from the standard curve.

TABLE 2. Carbohydrate standard dilution protocol

Tube No.	Glucose concentration (μg/ml)	Microliters of glucose stock soln.	Microliters of distilled water
1	100	500	0
2	75	375	125
3	50	250	250
4	25	125	375
5	10	50	450
6	0 (blank)	0	500

References

1. Geesey, G.G., R. Mutch, J.W. Costerton, and R.B. Green. 1978. Sessile bacteria: an important component of the microbial population in small mountain streams. Limnol. Oceanogr. 23:1214-1223.

2. Strickland, J.D. and T.R. Parsons. 1968. A practical handbook for seawater analysis. Bull. Fish. Res. Bd. Can. 167.

CHAPTER 13. METHODS FOR EVALUATION OF MICROBIAL-INFLUENCED CORROSION

Visual Examination

Often, visual examination of a surface can provide clues to the types of reactions that are responsible for an observed corrosion event. Certain surface modifications have been attributed to the presence of specific bacterial types. Characteristic deposits may serve as indications of microbially influenced corrosion (MIC). Discrete mounds and tubercles on surfaces often contain populations of bacteria.

Biologically-formed deposits are usually soft and easily deformed. Soft brown deposits are characteristic of MIC of 304 stainless steel in cooling water systems [4]. Bacterial analysis showed that the bulky deposits which formed on a flange (Plates 1 and 2) contained high densities of slime-producing pseudomonads, a few filamentous iron bacteria resembling *Sphaerotilus*, and sulfate-reducing *Desulfovibrio* in the black underlayer. Sulfate-reducing bacteria often produce gouge-like indentations on stainless steels. The gouges may be viewed after removal of the overlying deposit (Figure 1).

Crevice corrosion of stainless steel can also be related to the presence of microbes. Crevices often occur near but not in contact with slime deposits like those shown in Figure 1 (Figure 2). Unlike the shiny appearance of the gouge-type irregularity, crevices are often coated with a black deposit of FeS [5].

Irregularly distributed pits on alloys such as stainless steels are also possible sites of MIC. *Gallionella* on stainless steels produce subsurface cavities yielding only pinhole-sized penetrations at the surface (Plate 3). Pitting corrosion may also be found under biofilms on copper heat exchanger tubing (Plate 4). Numerous types of slime-producing bacteria were isolated from a pinhole pit in the

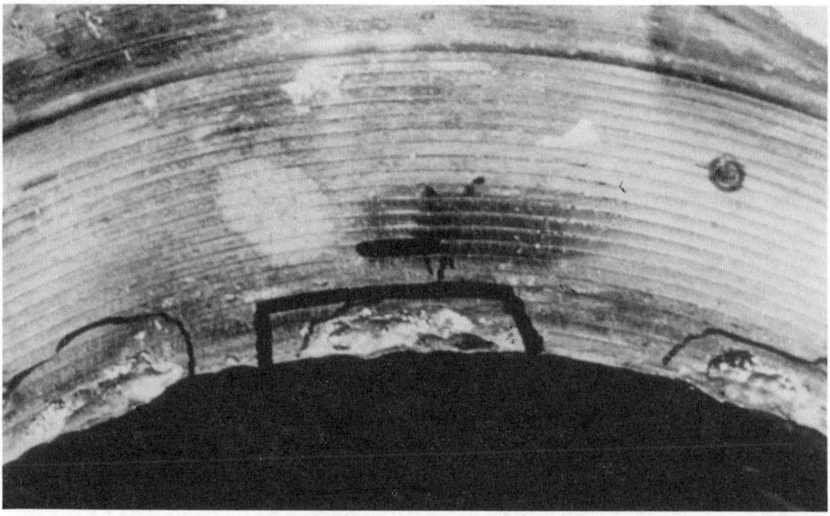

FIGURE 1. Closeup view of the surface under the biodeposit shown in Plate 1 after removal of the corrosion products. Shiny, active gouges were observed immediately beneath a black deposit that was composed mostly of chromium and sulfur. Courtesy of R.E. Tatnall [5] and NACE Publications, Houston, TX

tubing (Geesey, unpublished results).

Evidence of deposits can even be seen under coatings applied to metal surfaces. A carbon steel heat exchanger head liner with a bitumenous coating gave rise to numerous irregularly-distributed tubercles under the coating (Figure 3). A clear liquid was trapped within the deposit. The deposit itself exhibited a black coloration underneath. Many of the tubercles seen in Figure 3 covered rounded pits that completely penetrated the 5/8 in.-thick vessel wall. Surface color may also provide information on the source of corrosion. A black coloration similar to that shown in Plate 3 is indicative of an iron sulfide deposit formed as a result of sulfate-reducing bacterial activities. An orange or brown deposit suggests the presence of iron-oxidizing bacteria.

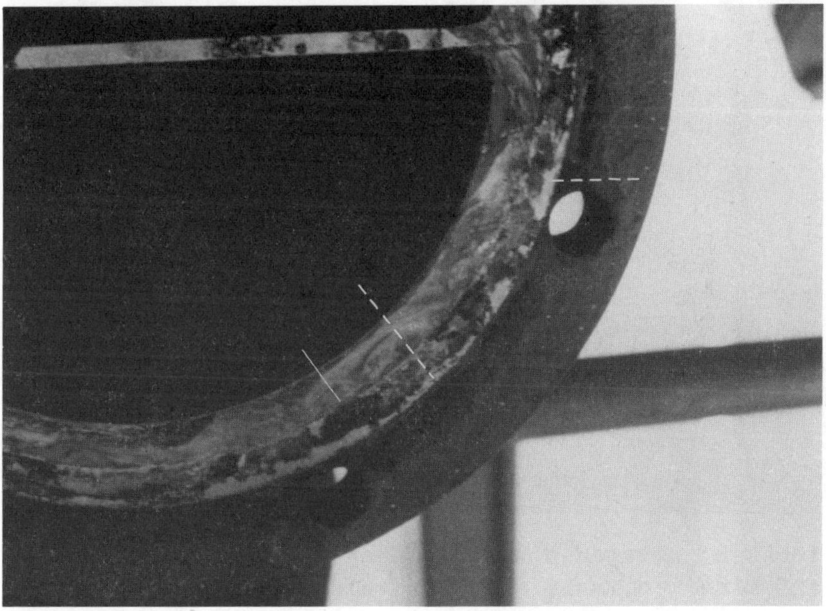

FIGURE 2. 304 stainless steel flange face showing corrosion which occurred under the gasket but always adjacent to, rather than directly beneath the biodeposit. Some corroded areas are filled with black deposits that liberate hydrogen sulfide when treated with hydrochloric acid (solid white pointer). Corroded areas with black deposits are interspersed with bright, shiny, corroded areas with no corrosion products (broken white pointer). Courtesy of R.E. Tatnall [5] and NACE Publications, Houston, TX

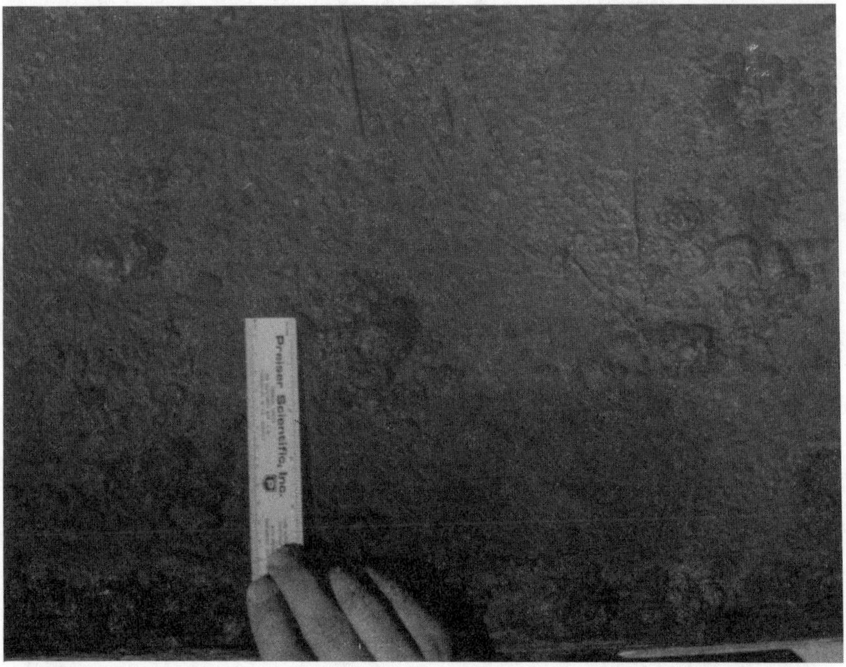

FIGURE 3. Bitumen-coated, carbon steel heat exchanger head liner commonly used for water service. Note the discrete tubercules, many of which appear to be growing under the intact coating. Corrosion products consist mostly of iron, phosphorus, and sulfur. When the coating is removed around the tubercle, a black deposit is exposed and a clear liquid oozes from the area. Many of the tubercles in this photo are covering rounded pits that have completely penetrated the 5/8 in.-thick vessel wall. Sulfate-reducing bacteria were isolated from the tubercles. Courtesy of R.E. Tatnall [6] and NACE Publications, Houston, TX

Gallionella may produce a reddish deposit on stainless steels. The growth of slime-producing bacteria will lead to the entrapment of suspended particles from the water so that the surface assumes the color of the particles.

Visual examination should always be followed by chemical and biological analysis of the corrosion deposit or pit area. Correct interpretations depend heavily on the sampling procedure that is employed. The investigator is advised to take the time to consider carefully a strategy that optimizes recovery of the problem-causing organism(s). It should be kept in mind that there are likely several types of microorganisms present which, through cooperative efforts, influence the corrosion reaction. The following sections should provide some useful procedures for microbiological sampling and enumeration. By no means do they apply to all types of MIC and consultation with a trained microbiologist familiar with corrosion-causing microorganisms is recommended if a successful strategy cannot be developed by in-house personnel.

Coupon Testing

Coupon substrates are commonly used to measure corrosion rates of metals submerged in aqueous systems [3]. Coupons are strips of metal composed of an alloy that resembles the material in a system that is subject to corrosion. A typical coupon has dimensions of 50 x 25 x 1 mm with a total exposed surface area of approximately 25 cm^2. The coupons may be fabricated in house to meet specific requirements of the user or obtained commercially from one of a variety of suppliers of corrosion monitoring equipment. The coupons are prepared for a test by the user. This generally involves the following steps: 1) sandblasting 2) mechanical polishing with 600 grit cloth or steel wool 3) degreasing with trichloroethylene 4) rinsing with distilled water 5) drying 6) weighing to 4 or 5 significant figures using an analytical balance before introduction into the water system.

The coupons should be oriented with respect to water flow in a manner that mimics the corroding metal surfaces in the system. The edges of the coupon should be masked since they generally exhibit a higher corrosion current density than the other exposed surfaces of the coupon. A sufficient number of coupons should be introduced into the system so that different exposure periods can be tested. Sampling of replicate coupons at any one time does not appear to provide any advantage over sampling a single coupon since variations are great, even when as many as 30 replicates are taken.

Estimation of Corrosion Rate

In the case of corrosion-causing biofilms, the rate of weight loss of a coupon is directly related to the corrosion rate of the metal surface under the biofilm and indirectly related to the effectiveness of a particular biocide treatment. This evaluation can be performed with only a modestly equipped laboratory and a technical staff with minimal training.

After a coupon has been exposed to the water for a period of time sufficient to cause measureable weight loss, it is removed from the system and dried, taking care not to dislodge loosely-bound corrosion products, and then reweighed. Visual examination should follow weight loss determinations. The general appearance of the surface should be noted. Pit density and deposit hardness are among the parameters that should be considered. Microscopic examination should then be carried out to estimate corrosion depth, path of corrosion, and any other changes that may have occurred at the microscopic level.

Corrosion rate (C.R.), based on the Metric System is expressed as milligrams of metal lost/square decimeter/day (mdd) and is calculated by:

$$\text{C.R.} = \frac{\text{weight change (milligrams)}}{\text{area of coupon (dm}^2\text{) x days of exposure}}$$

To convert mdd to another common C.R. expression (mils/yr), multiply mdd by 1.437/D, where D is the density of the metal. The density of some common metals are given below.

TABLE 1. Densities of some common metals

Metal	g/cc
Aluminum	2.70
Copper	8.96
Iron	7.87
304 Stainless Steel	7.92
Magnesium	1.74
Titanium	4.5

Detection of Corrosion-Causing Microorganisms Using Cultural Methods

Microorganisms from diverse physiological groups have been implicated in corrosion-causing reactions at metallic surfaces. Sulfur-oxidizing bacteria such as thiobacilli are capable of oxidizing $S_2O_3^{2-}$, S^o, $S_4O_6^{2-}$, SCN, and H_2S to H_2SO_4. The acid by-product exerts a corrosive effect on many metals. Sulfate-reducing bacteria such as those of the genera *Desulfovibrio* and *Desulfotomaculum* reduce SO_4^{2-} to S^{2-}. Iron bacteria such as *Gallionella* and *Sphaerotilus* promote corrosion of steel by oxidizing ferrous iron to ferric iron. Slime-producing, aerobic bacteria such as *Pseudomonas* spp. promote corrosion by creating anaerobic niches in biofilms covering metal surfaces which are suitable for growth of anaerobic sulfate-reducing bacteria. They also facilitate the establishment of oxygen and metal concentration cells across the surface.

Microorganisms that fall into these basic physiological groups possess a diverse set of nutrient requirements which must be satisfied before they can be cultured. Growth is a prerequisite for detection and enumeration of microorganisms based on the culture methodology. Consequently, no one or even several types of culture medium supports the growth of all potential corrosion-causing microorganisms. The more that is known about the conditions in the environment where the corrosion event is taking place, however, the greater will be the chance of selecting or formulating a culture medium that enables recovery of the corrosion-causing microorganism(s). The investigator is therefore encouraged to determine as thoroughly as possible the physical and chemical conditions that characterize the system before attempting to enumerate or isolate the microorganisms present.

Years of experience have produced several "standard" media formulations that satisfy the nutrient reqirements of several bacterial types that are commonly associated with MIC. Enumeration procedures have been developed which, in some cases, provide bacterial densities that seem to correlate with incidence and extent of some types of corrosion processes. These may be useful in assessing MIC in some systems. Additional media for the cultivation of microorganisms involved in MIC may be obtained from Weber [7].

Media Preparation

Directions for the preparation of media that

have been reported to support the growth of various bacteria associated with corrosion of metallic surfaces may be found in the appendices at the end of this chapter.

Sample Collection

1. Collection material and vessels should be sterile. This is achieved by autoclaving at 121°C for 20 min.

2. Particulate samples (i.e., deposit or slime material) should be fresh and kept in a hydrated state during transport and storage. Source water from the system should be used. An effort should be made to preserve the structural integrity of the sampled material. Samples collected from anaerobic environments should be maintained in an anaerobic state using an N_2 gas overlay or other appropriate means.

3. Although storage of samples prior to microbiological processing is not recommended, culturing procedures should be performed in a properly outfitted laboratory in order to avoid sample contamination. All manipulations should be performed in a certified biological hood or anaerobic glove box whenever possible. Quantitative evaluation of microorganisms usually requires that the sample be processed within 12 h of collection. Samples obtained for qualitative evaluation should be processed within 72 h of collection.

4. Different areas of a sampled deposit or slime layer should be aseptically dissected wherever heterogeneity is observed. This will provide the opportunity to demonstrate any community structure that may exist as well as allow

isolation and enumeration of important microbial populations other than the dominant species.

5. When estimates of bacterial densities are desired, each dissected sample fraction should be homogenized separately in a sterile micro-blender assembly in order to disrupt clumps and aggregates of bacteria. Particulate samples should be suspended in a solution composed of filter-sterilized water obtained from the system containing the deposit. For anaerobic samples, blending should be performed in an anaerobic chamber. Blending speed and time should be adjusted to achieve maximum dispersion of microorganisms with minimum cell disruption. Usually 30-60 seconds at 60% maximum speed is adequate.

6. It should be noted that sampling the bulk liquid (water) phase generally provides little useful information with regard to the microorganisms responsible for surface corrosion activities, even when that surface is in direct contact with the liquid.

Enumeration of anaerobic microorganisms (dilution to extinction)

1. A volume of blended sample equivalent to 1/10th that of the anaerobic, sterile medium contained within a series of dilution tubes is drawn into a 1 or 2 ml sterile syringe and injected through the serum stopper into the first of 5 tubes containing the appropriate medium. Care should be taken to avoid introduction of air during sampling and inoculation of tubes when enumerating microorganisms from anaerobic environments.

2. The contents of the tube are mixed thoroughly

with a vortex mixer and a volume equivalent to 1/10th the liquid volume in the tube is transferred with a new sterile syringe to another tube of sterile anaerobic medium and the contents are mixed as above.

3. The procedure is repeated with each newly inoculated tube so that a 10-fold serial dilution is achieved with each transfer until all 5 tubes have been inoculated. This should provide a 1 to 100,000 dilution of the original sample in the last tube.

4. This 5 tube inoculation procedure is repeated 2 more times with fresh subsamples from the blended suspension and sterile tubes of medium so that triplicate dilutions are prepared for each sample.

5. The tubes are placed in an incubator set at a temperature which reflects that of the environment from which the sample was obtained and incubated for a period of time that will allow the growth of any microorganisms present in the tubes. Although the incubation period depends on the growth rate of the microorganisms present, no more than 3 weeks is generally required to obtain visible evidence of growth.

6. Tubes showing any increase in turbidity or precipitate after a period of incubation greater than that of uninoculated control tubes are scored as containing at least 1 viable microorganism. The number of microorganisms in the original, undiluted sample may be estimated from the tube with the highest dilution giving evidence of growth. The 3 replicates should be within 1 dilution factor of each other.

Example:

Growth occurred in the first 3 dilution tubes of 1 dilution series and in the first 4 tubes of the 2 remaining dilution series. Thus, there were at least 100 but not more than 1000 microorganisms in the original subsample volume used to inoculate the first dilution series and at least 1000 but not more than 10,000 microorganisms in the original subsamples used to inoculate the other 2 dilution series. Although this enumeration procedure is not precise, it gives a gross indication of the density of viable microorganisms that replicate in a particular medium under a specified set of conditions. It is also possible to apply the Most Probable Number method of enumeration to obtain more accurate estimates of the density of bacteria in the sample. The reader is referred to the Standard Methods for the Examination of Water and Wastewater reference manual [1] for details on this approach.

Enumeration of aerobic microorganisms

The procedure for enumeration of microorganisms that require the presence of oxygen for growth is similar and somewhat easier to perform than that described above for anaerobic microorganisms. Media can be distributed into screw cap tubes with no precautions taken to exclude air. Sample blending and transfer need not be conducted in an anaerobic environment.

References

1. American Public Health Association. 1985. Standard methods for the examination of water and wastewater. 16th ed. American Public Health Association, Washington, D.C.

2. Kobrin, G. 1986. Reflections on microbiologically induced corrosion of stainless steels. p. 33-46, In S.C. Dexter (ed.), Biologically induced corrosion, NACE Publications, Houston, TX

3. NACE Basic Corrosion Course. 1982. Chapter 15, NACE Publications, Houston, TX.

4. Tatnall, R.E. 1981a. Fundamentals of bacteria induced corrosion. Materials Performance 20:32-38.

5. Tatnall, R.E. 1981b. Case histories: bacteria induced corrosion. Materials Performance 20:41-48.

6. Tatnall, R.E. 1984. Experiments in microbiologically influenced corrosion. Paper No. 95, National Association of Corrosion Engineers annual meeting, NACE Publications, Houston, TX.

7. Weber, G.R. 1965. A study for the development of procedures for isolation of chemoautotrophic and related microorganisms and the characterization of such microorganisms from corroded metals. Quarterly Progress Report no. 3, NASA contract no. NASW 1017, National Distillers and Chemical Corp., 49p.

PLATE 1. Cooling tower condenser head flange composed of 304 stainless steel that was covered with an asbestos gasket and a stainless steel blind flange. After 3 years exposure to untreated river water, evidence of severe crevice corrosion appeared in and near flanged and gasketed joints. The corrosion was characterized by tan to brown slimy deposits from which slime-forming and iron bacteria were isolated (arrow). Courtesy of R.E. Tatnall [4] and NACE Publications, Houston, TX.

PLATE 2. String of bio-deposits (arrows) along a weld seam in a head flange made of 304 stainless steel. Courtesy of G. Korbin [2] and NACE Publications, Houston, TX.

Plate 3. Closeup view of a mound-like bio-deposit on austenitic 304L stainless steel tank at a butt-welded seam still wet with test water. Deposit was slimy and gelatinous in texture and measured 3-4 in. in diameter. Pits in and adjacent to welds under the reddish-brown deposits were detected after several months exposure to untreated well water. Large, deep, bottle-shaped pits were only revealed after probing deposit at the edge of the weld seam with an ice pick (not shown). High counts of _Gallionella_ and _Siderocapsa_ were recovered from the deposit. Courtesy of G. Kobrin [2] and NACE Publications, Houston, TX.

Plate 4. Copper tubing from heat exchanger showing corrosion products on water-exposed surface after removal of overlying biological slime layer. Pinhole-sized pits that extend completely through the tubing walls were

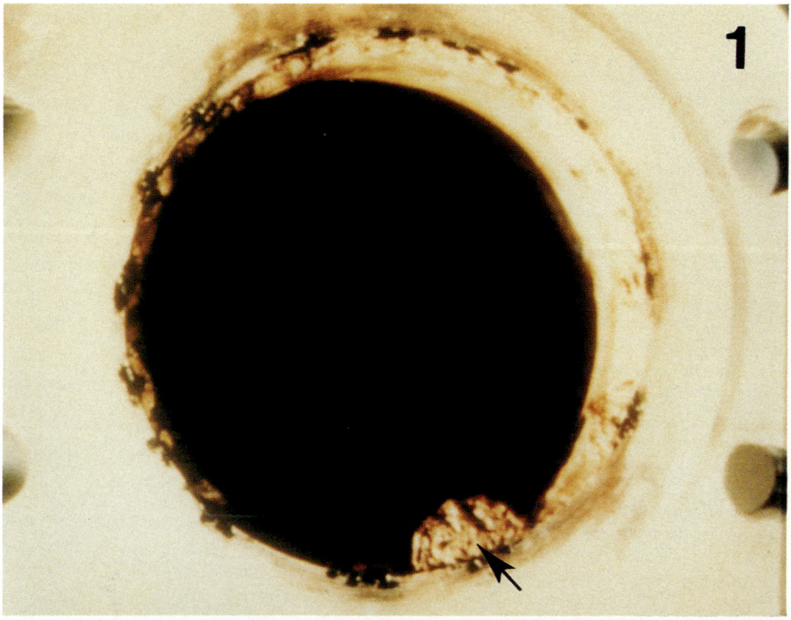
1

2

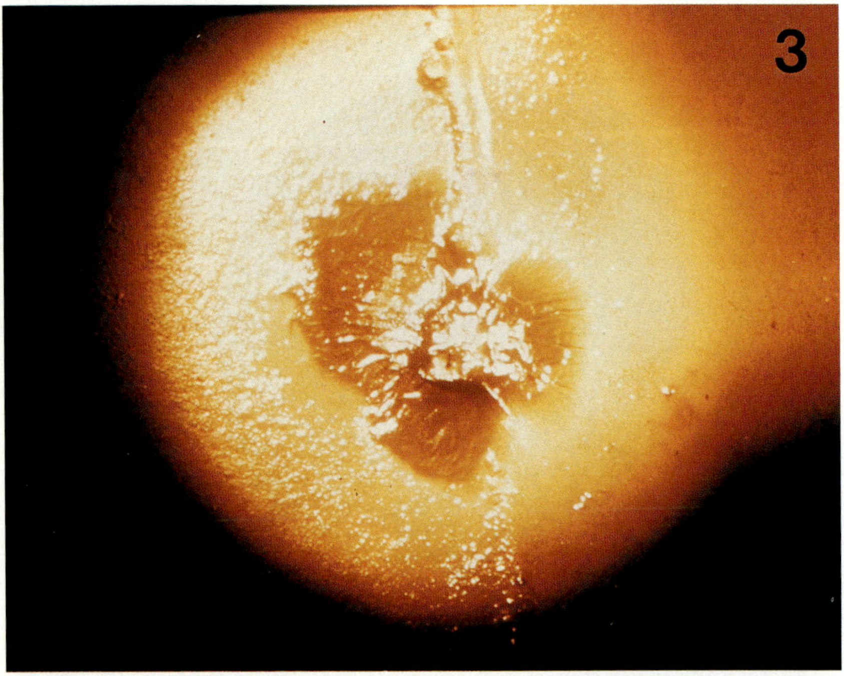
3

4

observed beneath the corrosion products (arrows). Courtesy of W.F. Mayer, Aqua-Serv Engineers, Inc.

Appendices

Appendix 1 Sulfur-oxidizing bacteria (Thiobacillus)

Sulfur Oxidizing Bacteria Medium

$(NH_4)_2SO_4$	0.2 to 0.4 g
KH_2PO_4	3 to 4 g
$CaCl_2$ (filter-sterilized)	1.25 g
$MgSO_4 \cdot 7H_2O$	0.5 g
$FeSO_4$	10 mg
elemental sulfur	10 g
tapwater	1 liter

Adjust pH to 5.0. Sterilize by steaming for 30 min. The $CaCl_2$ is added after the medium is autoclaved and cooled to room temperature. Growth of sulfur oxidizing bacteria is indicated by a dramatic reduction in medium pH, 4 to 14 days after inoculation. An abrupt drop to pH 2.5 or lower indicates growth of *Thiobacillus thiooxidans*. Little turbidity develops in the medium. Thus, for enumeration purposes, a low pH rather than growth (turbidity) would be scored in various dilution tubes.

Reference

1. Shreir, L.L. 1963. The microbiology of corrosion. p. 2.51-2.64. *In* Corrosion, Vol. 1, J. Wiley and Sons, New York.

Appendix 2 Sulfate-reducing bacteria

Butlin's Medium

sodium lactate (60%)	3.5 ml
K_2HPO_4	0.5 g

NH_4Cl	1.0 g
Na_2SO_4	2.0 g
$MgSO_4 \cdot 7H_2O$	1.0 g
yeast extract	1.0 g
$CaCl_2 \cdot 6H_2O$	0.1 g
distilled water	1 liter
resazurin (0.1% soln.)	1 ml

The medium is boiled in a stoppered flask for approximately 40 min during which time a stream of N_2 gas is bubbled into the solution until the medium changes from a yellow to a pink/orange color and finally to a straw yellow color. Water may be added periodically to make up for that lost during boiling. Five ml of the medium are then transferred from the flask to tubes with a Cornwall dispensing syringe. The tubes should contain an uncoated nail and be purged with a stream of N_2 gas just prior to and during addition of the medium. A tight-fitting, neoprene serum stopper is used to seal the tubes immediately after addition of the medium. The tubes are secured in a rack containing a clamping top-plate that holds the stoppers in place during autoclaving. The tubes should not develop any pink color after autoclaving.

The nail will turn black if sulfate-reducing bacteria develop after inoculation. Cultures should be maintained under anaerobic conditions.

API Medium RP-38

sodium lactate	4.0 g
yeast extract	1.0 g
ascorbic acid	0.1 g
$MgSO_4 \cdot 7H_2O$	0.2 g

K_2HPO_4	0.01 g
$Fe(NH_4)_2(SO_4)_2 \cdot H_2O$	0.1 g
NaCl (if sample is saline)	10.0 g
distilled water	1 liter

The medium pH is adjusted so that after autoclaving (121°C, 20 min), it is between 7.0 and 7.5.

The ingredients are dissolved by gentle heating. Any medium which develops a heavy precipitate should be discarded. The medium is dispensed in tubes, deaerated by bubbling with N_2 gas, and stoppered with neoprene serum stoppers and autoclaved in a rack as described for the Butlin's Medium. The nail will turn black if sulfate-reducing bacteria develop after inoculation. Cultures should be maintained under anaerobic conditions.

References

1. The role of bacteria in the corrosion of oil field equipment. 1982. TPC3 Publication, National Association of Corrosion Engineers, Houston, TX.

2. API recommended practice for biological analysis of injection water. 1959, American Petroleum Insititute (API) RP38.

Postgate's Medium B

KH_2PO_4	0.5 g
NH_4Cl	1.0 g
$CaSO_4$	1.0 g
$MgSO_4 \cdot 7H_2O$	2.0 g

sodium lactate (60%)	3.5 g
yeast extract	1.0 g
ascorbic acid	0.1 g
thioglycollic acid	0.1 g
$FeSO_4 \cdot 7H_2O$	0.5 g
Tap water	1 liter

Adjust pH to between 7.0 and 7.5. This medium always contains a precipitate. NaCl should be added for isolation of bacteria from saline environments. The medium should be prepared and autoclaved just before inoculation since the reductants deteriorate in air at neutral pH, a process which is accompanied by the development of a purple coloration. A black precipitate indicates that SRB are present.

Reference

1. Postgate, J.R. 1984. The sulfate-reducing bacteria. Cambridge University Press, New York. 208 p.

Appendix 3 Sulfite-Reducing Bacteria

Butlin's Medium is prepared as described above except that the sodium sulfate is replaced with an equivalent amount of deoxygenated, filter-sterilized sodium sulfite after the medium has been autoclaved. Cultures should be maintained under anaerobic conditions.

Appendix 4 Iron-Reducing Bacteria

B_{10} Medium

K_2HPO_4	0.8 g
KH_2PO_4	0.2 g

$MgSO_4 \cdot 7H_2O$	0.2 g
NaCl	0.2 g
$MnSO_4 \cdot 2H_2O$	0.0001 g
$NaMoO_4$	0.0001 g
$CaSO_4$ (saturated soln.)	10 ml
yeast extract	5.0 g
peptone (Difco)	5.0 g
soluble $FePO_4$	4.7 g
agar	15.0 g
distilled water	990 ml

Adjust pH to 7.2-7.4 with 0.1 M HCl

Incubate plates for 4-7 days at appropriate temperature. Colonies with concave morphology and light orange color are characteristic of iron-reducing bacteria. Liquid medium (without agar) turns from a deep brown (ferric ions) to a green color (ferrous ions) in the presence of iron reducing bacteria. Cultures should be maintained under aerobic conditions.

Reference

1. Obuekwe, C.O., D.W.S. Westlake, and F.D. Cook. 1981. Effect of nitrate on reduction of ferric iron by a bacterium isolated from crude oil. Can. J. Microbiol. 27:692-697.

<u>Appendix 5</u> Iron Oxidizing Bacteria (*Sphaerotilus*)

glucose	1.0 g
peptone	1.0 g
$MgSO_4 \cdot 7H_2O$	0.2 g
$CaCl_2$	0.05 g

$FeCl_3 \cdot 6H_2O$	0.01 g
tapwater	1 liter

Adjust pH to between 7.0 and 7.5.

Reference

1. The Procaryotes. 1981. Vol. 1, M.P. Starr, H. Stolp, H.G. Truper, A. Balows, and H.G. Schlegel (ed.), Springer-Verlag, New York.

Appendix 6 Iron Oxidizing Bacteria (*Thiobacillus*)

Basal salts

$(NH_4)_2SO_4$	0.15 g
KCl	0.15 g
K_2HPO_4	0.15 g
$MgSO_4 \cdot 7H_2O$	3.36 g
$CaCl_2$	0.97 g
$Al_2(SO_4)_3 \cdot H_2O$	2.25 g
$MnSO_4 \cdot H_2O$	0.12 g
$FeSO_4 \cdot 7H_2O$	1.5 g
distilled water	1 liter

The basal salts solution and ferrous iron solutions are made up as 2X concentrated solutions in 500 ml volumes, the pH of each is adjusted, and then each solution is autoclaved separately, and cooled before being combined and distributed aseptically into sterile tubes. Preparation of this medium results in a cloudy mixture. Following inoculation, the tubes are incubated for 3 weeks at a temperature comparable to that of the system from which the inoculum was obtained. Growth of iron oxidizing bacteria is indicated by the development of a red-orange color in the tubes.

Reference

1. Unz, R.F. and T. Lieberman. 1973. The microbiology of acid mine water treatment in packed bed columns. Research Publication No. 77, The Pennsylvania State University, University Park, PA, 73 p.

Appendix 7 Slime-Producing Pseudomonads

Pseudosel Agar (BBL no. 11554)
Pseudomonas Isolation Agar (Difco no. 0927-01)

Reference

1. Pope, D.H. 1986. Discussion of methods for the detection of microorganisms involved in microbially influenced corrosion. p. 275-283. In S.C. Dexter (ed.), Biologically induced corrosion. NACE Publications, Houston, TX.

Appendix 8 Total Aerobic Bacteria

Plate Count Agar (Difco no. 0479-01)

Appendix 9 Total Anaerobic Bacteria

Brucella Broth (Difco no. 0495-01)

CHAPTER 14. BIOCIDE TREATMENT EVALUATIONS

Throughout this book, the significance of sessile microbial populations as initiators of biofouling activities has been emphasized. The efficacy of chemical (and physical) treatments is therefore best assessed when these populations--rather than planktonic organisms--are used to challenge various biocides. The use of coupons and/or biofilm sampling devices such as the Robbins Biofilm Sampler (insert on inside book cover) are useful tools for this type of testing.

The first section of this chapter presents a method for the more traditional, planktonic bottle test assay. Data from this type of planktonic assay is, however, useful in making a determination of biocide compatibility with the industrial water under consideration. Should a particular treatment compound prove effective using the planktonic assay, the methods discussed in the second half of this chapter can be employed to determine sessile population removal efficacies. References have been provided at the end of the chapter that discuss biocide testing and the effects of biocides on biofouling microorganisms [1-9].

Biocide Efficacy Testing: Planktonic Bottle Assay

Materials & Reagents

1. Sample bottles, 100 ml borosilicate milk dilution with screw-top caps.

2. Analytical balance, sensitivity to 0.1 mg with glass weighing boats.

3. Constant temperature water bath, precision to

within 1°C.

4. Sterile pipets & petri dishes.

5. Autoclave or dry heat sterilizer.

6. Magnetic stirrer.

7. Volumetric flasks, 100-1,000 ml.

8. Membrane filtration apparatus, Millipore or equivalent.

9. Distilled/purified water, reagent-grade.

10. Tryptone glucose yeast extract agar (TGYE), DIFCO or equivalent.

11. Sabouraud dextrose agar, DIFCO or equivalent.

12. 0.45 μm sterile membrane filters, Nuclepore or equivalent.

13. 0.1% peptone broth, sterile, DIFCO or equivalent.

14. Appropriate neutralizing agents for specific biocides; e.g., sodium thiosulfate solution for chlorine, sterile dilution blanks.

15. Incubator, constant temperature.

16. Colony counter, Quebec or equivalent.

Preparation of Biocide Solutions

1. Prepare appropriate dilutions (typically, 1-100 mg/l) of the test compounds in reagent grade

water or other reagent grade solvents required for dissolution. A minimum of 6 concentrations should be tested for each compound. The volume of biocide dosed into the challenge water should not exceed 1-2% of the total volume.

2. Store the prepared stock dilutions in pre-cleaned, glass volumetric flasks.

Test Procedure

1. Select biocide contact times appropriate for the compounds under consideration; biocides differ in their modes of action and are often time-dependent insofar as efficacy is concerned. Contact intervals of 30 minutes up to 3 or more hours are recommended as a starting point.

2. During biocide contact, the test solution should be maintained at the same temperature as that of the water system under consideration.

3. Place the test water containing indigenous microbial populations in borosilicate glass milk dilution bottles or other appropriate containers and place the containers in the constant temperature bath for equilibration. A minimum of 10^5 cells/ml should be present in the sample; artificially contaminated waters can also be prepared by addition of the appropriate pure or mixed cultures to the water under consideration. Prior to addition of the biocide, aliquots of the sample water should be removed and processed via steps 6 and 7, below. These samples will serve as T=0 min. controls.

4. The assay starting times should be staggered (2-3 min. per biocide dilution) to allow sufficient

time for sample processing.

5. At the appropriate time intervals, add a minimum volume of the biocide stock solution such that the desired concentration is obtained.

   For example, if a 10 ppm biocide concentration is desired, 1 ml of a 1,000 ppm stock solution should be added to 99 ml of sample. The test solution should be stirred continuously during the contact period.

6. At the end of each contact period, remove aliquots from the test solution to dilution blanks containing the appropriate neutralizing solutions. Gently agitate the neutralization solutions for a minimum of 1 min. Dilutions of between 10^{-1} and 10^{-5} are recommended.

7. Filter a minimum of 5 ml from each dilution through a 0.45 μm sterile membrane filter. Rinse each filter with two-50 ml volumes of peptone broth, then aseptically transfer the filter to the appropriate agar (TGYE for bacteria; Sabouraud dextrose for fungi). For purified water applications, 0.3% TSA can be substituted for the TGYE agar. Incubate bacterial membranes for 48-96 h at 30-35°C; fungal membranes, 96-120 h at 25-30°C.

8. Following incubation, colonies are counted using a colony counter. Results are calculated by dividing the number of colonies observed by the appropriate dilution factor.

Data Evaluation

1. Control samples should elicit either a stable or

increasing number of cells during the contact period. Results should be evaluated by determining the number of organisms killed at each biocide concentration for the various contact times relative to the control counts.

2. Effectiveness may be defined based upon a 90% (1 log) reduction in viable numbers.

3. That dilution producing a 90% reduction in viable numbers relative to the control value at T= x min should be considered for further testing by coupon or other biofilm sampling assays.

Biocide Efficacy Testing: Sessile Assays

Coupons such as those described in Chapter 11 are convenient to use for evaluation of treatments that are applied to the water to control corrosion and biofilm development. Biocides and corrosion inhibitors can be tested for their effectiveness in controlling the rate of biofilm development in a system. Tests can be performed "in-line" or with a side-stream device. In the case of slug dose applications, fouled coupons are removed from in-line or side-stream devices and placed in biocide-containing process water to approximate conditions during an actual biocide application. Laboratory evaluations of biocides can be improved by the use of coupons that are initially colonized by microorganisms in the system. This usually involves working with mixed populations of microorganisms. Evaluation of coupon fouling in the presence of biocides may be carried out by scanning electron microscopy, epifluorescence microscopy, or by cultural methods.

Biocide Treatment Evaluation

Cultural Method

1. Identical coupons of appropriate composition are place in untreated and biocide-treated process water for a sufficient period of time to enable evaluation of biocide efficacy.

2. Each coupon is recovered aseptically from the test solutions and rinsed with sterile process water.

3. The biofilm is removed quantitatively from each coupon with a sterile scalpel blade and transferred to a separate tube containing 10 ml of sterile process water or distilled water.

4. Each tube is placed in a water bath sonicator for 1 min to disperse the cells.

5. The contents of each tube are serially diluted and 0.1 ml of each dilution is spread on the surface of solid growth medium composed of 0.3% Trypticase Soy Agar in a petri dish. Another medium may be substituted in order to encourage the growth of the test microorganisms.

6. The dishes are inverted and incubated at the process water temperature until visible colonies appear.

7. Enumerate the colonies appearing on the dishes that contain 30-300 colonies and note the dilution used for the counts.

8. Calculate the number of viable organisms on the coupon using the following formula:

Biocide Treatment Evaluation

No. viable organisms = colonies/plate x dilution factor x 10

Example:

Coupon A (no biocide treatment)	220 colonies on 10^{-6} plate
Coupon B (biocide-treated)	44 colonies on 10^{-2} plate

Coupon A $220 \times 10^6 \times 10 = 2.2 \times 10^9$ cfu/coupon

Coupon B $44 \times 10^2 \times 10 = 4.4 \times 10^4$ cfu/coupon

Percent reduction = $1 - (4.4 \times 10^4 / 2.2 \times 10^9) \times 100$

Coupons can also be removed at timed intervals to produce a "D-value" which is the time required to reduce the bacterial concentration on the treated coupon to 10% of that obtained for the untreated coupon.

References

1. American Society for Testing and Materials. 1978. Efficacy of microbicides used in cooling systems. ASTM document E645-78. ASTM, Washington, D.C.

2. Block, S. 1983. Sterilization, disinfection, and preservation. Lee and Febiger, Philadelphia.

3. Characklis, W.G., N. Zelver, R.O. Lewis, C.H. Nelson, D.E. Dobb, and G.K. Pogenkopf. 1983. Influence of biofouling and biofouling control techniques on corrosion of copper-nickel tubes.

Paper no. 250. Proceedings of the National Association of Corrosion Engineers annual meeting, NACE Publications, Houston, TX.

4. Eagar, R.G, A.B. Theis, M.H. Turakhia, and W.G. Characklis. 1986. Glutaraldehyde: impact on corrosion causing biofilms. Paper no. 125. Proceedings of the National Association of Corrosion Engineers annual meeting, NACE Publications, Houston, TX.

5. Kajdasz, R., R.V. Einstman, and L. Young Bandala. 1984. Biocide efficacy with respect to sessile and planktonic organisms. p. 85-93. Proceedings of the international water conference annual meeting, Pittsburg, PA.

6. McCoy, J.W. 1980. The Microbiology of cooling water. Chemical Publishing Co., NY.

7. McCoy, W.F., J.E. Ridge, and E.S. Lashen. 1986. Efficacy of biocides in laboratory model cooling towers. Materials Performance 25:9-14.

8. Ruseska, I., J. Robbins, J.W. Costerton, and E.S. Lashen. 1982. Biocide testing against corrosion-causing oil-field bacteria helps control plugging. Oil & Gas J. 80:253-264.

9. Strauss, S.D. and P.R. Puckorius. 1984. Cooling-water treatment for control of scaling, fouling, corrosion. Power 128(6)S1-24.

Index

Index

Index